# HIGH-RESOLUTION
# XAS/XES

## Analyzing
## Electronic Structures
## of Catalysts

# HIGH-RESOLUTION
# XAS/XES

## Analyzing Electronic Structures of Catalysts

**Edited by**

## Jacinto Sá

CRC Press
Taylor & Francis Group
Boca Raton London New York

CRC Press is an imprint of the
Taylor & Francis Group, an **informa** business

CRC Press
Taylor & Francis Group
6000 Broken Sound Parkway NW, Suite 300
Boca Raton, FL 33487-2742

First issued in paperback 2017

© 2015 by Taylor & Francis Group, LLC
CRC Press is an imprint of Taylor & Francis Group, an Informa business

No claim to original U.S. Government works

Version Date: 20140514

ISBN 13: 978-1-4665-9298-8 (hbk)
ISBN 13: 978-1-138-07157-5 (pbk)

---

**Library of Congress Cataloging-in-Publication Data**

---

High-resolution XAS/XES : analyzing electronic structures of catalysts / [edited by] Jacinto Sa.
    pages cm
Includes bibliographical references and index.
    ISBN 978-1-4665-9298-8 (hardback)
    1. Emission spectroscopy. 2. Absorption spectra. 3. Catalysts. 4. Enzymes--Electric properties. I. Sa, Jacinto, editor.

QD96.E46H54 2014
541'.395--dc23                                                                            2014015949

---

**Visit the Taylor & Francis Web site at**
**http://www.taylorandfrancis.com**

**and the CRC Press Web site at**
**http://www.crcpress.com**

# Contents

# Contents

# Preface

Photon-in photon-out core level spectroscopy is an emerging approach to characterize the electronic structure of catalysts and enzymes, and it is either installed or planned for intense synchrotron beam lines (third generation of X-ray light) and X-ray free electron lasers (fourth generation of X-ray light). This type of spectroscopy requires high-energy resolution spectroscopy not only for the incoming X-ray beam but also, in most applications, for the detection of the outgoing photons. Thus, the use of high-resolution X-ray crystal spectrometers whose resolving power $\Delta E/E$ is typically about $10^{-4}$ is mandatory.

Herein, we have compiled the developments on X-ray light sources, detectors, crystal spectrometers, as well as the plethora of photon-in photon-out core level spectroscopy techniques that have been developed. The book also encompasses two sections on photon-in photon-out core level spectroscopy applications to the study of catalytic systems. The applications highlighted are restricted to hard X-ray measurements since they encompass most of the studies involving catalysis, primarily due to high penetration probes, enabling *in situ* studies.

The book is for undergraduate and graduate students, as well as for scientists working in the area of materials and catalysis. It should be mentioned that the goal of the book is to provide an overview of the field. The examples were chosen in order to cover a larger range of scientific experiments as concisely as possible. Therefore, and in the name of the authors, we would like to apologize for any work that has not been referenced or mentioned. Any omission was decided upon simply on the basis of chapter concision. Finally, the book was written in a language that we consider accessible to most undergraduate science students. Technical terms and mathematical expressions were mentioned only when strictly necessary.

I would like to finish on a personal note, and thank the chapter authors for their contributions and all the scientific works that made the execution of the book possible.

**Jacinto Sá, Ph.D.**
*Institute of Physical Chemistry*
*Polish Academy of Sciences*
*Warsaw, Poland*

# The Editor

**Jacinto Sá, Ph.D.,** (physical-chemistry) is the Modern Heterogeneous Catalysis (MoHCa) group leader at the Institute of Physical Chemistry, Polish Academy of Sciences, Warsaw, Poland. He received his M.Sc. (chemistry) in the field of analytical chemistry at the Universidade de Aveiro (Portugal), and did his research project at Vienna University of Technology (Austria). He was awarded a Ph.D. degree in 2007 by The University of Aberdeen (Scotland) in the field of catalysis and surface science. In 2007, Dr. Sá moved to the CenTACat group at Queen's University Belfast to begin his first postdoctoral fellowship under the guidance of Professors Robbie Burch and Chris Hardacre. During Dr. Sá's stay he was awarded an R&D 100 Award for his involvement in the development of SpaciMS equipment. In 2010, he moved to Switzerland to begin his second postdoctoral fellowship at ETH Zurich and the Paul Scherrer Institute. Dr. Sá's research efforts were focused on the adaptation of high-resolution X-ray techniques for the study of catalysts and nanomaterials under working conditions. In 2013, he joined the Laboratory of Ultrafast Spectroscopy, at École Polytechnique Fédérale de Lausanne (EPFL), Switzerland, to expand the use of high-resolution X-ray techniques into the ultrafast domain and to take advantage of the newly developed XFEL facilities.

Currently, Dr. Sá's research efforts are focused on understanding the elemental steps of catalysis, in particular, those taking place in artificial photosynthesis and nanocatalytic systems used in the production of fine- and pharma-chemicals. His experience in using accelerator-based light sources to diagnose the mechanisms by which important catalytic processes proceed, and more recently, in conventional ultrafast laser sources, makes him one of the most experienced researchers in the world in this area. He has had more than 70 publications in international scientific journals and more than 20 oral and 50 poster presentations at scientific congresses.

Dr. Sá is married to Cristina Paun and is expecting his first child, a baby boy to be named Lucca V. Sá. He is a member of the Portuguese think tank O Contraditorio, part of the English volunteer group of Red Cross Zurich, and a part-time DJ (DJ Sound it). He enjoys fine art, in particular impressionism and surrealism, traveling, music, cinema, and fine dining. (E-mail: jsa@ichf.edu.pl)

# Contributors

**Serena DeBeer**
Max Planck Institute for Chemical
   Energy Conversion
Muelheim an der Ruhr, Germany
and
Department of Chemistry and
   Chemical Biology
Cornell University
Ithaca, New York
serena.debeer@cec.mpg.de

**Jean-Claude Dousse, Ph.D.**
Department of Physics
University of Fribourg
Fribourg, Switzerland
jean-claude.dousse@unifr.ch

**Joanna Hoszowska, Ph.D., P.D.**
Department of Physics
University of Fribourg
Fribourg, Switzerland
joanna.hoszowska@unifr.ch

**Yves Kayser, Ph.D.**
Paul Scherrer Institute
Villigen-PSI, Switzerland
yves.kayser@psi.ch

**Christopher J. Milne, Ph.D.**
SwissFEL
Paul Scherrer Institute
Villigen-PSI, Switzerland
chris.milne@psi.ch

**Thomas J. Penfold, Ph.D.**
SwissFEL
Paul Scherrer Institute
Villigen-PSI, Switzerland
thomas.penfold@psi.ch

**Jakub Szlachetko, Ph.D.**
Paul Scherrer Institute
Villigen-PSI, Switzerland
jakub.szlachetko@psi.ch

# 1 X-ray Sources and Detectors

## Christopher J. Milne

## CONTENTS

## 1.1 INTRODUCTION

Since their initial discovery by Röntgen in 1895,[1,2] X-rays have represented a unique tool for investigation of matter. Their powerful combination of short wavelength and penetrating nature make X-rays ideal for performing structural measurements within bulk systems. Shortly after their discovery it was determined that X-rays are also sensitive to the elemental composition of a given sample, and X-ray absorption spectroscopy was born in the early 20th century.[3] The last century has seen a rapid development of X-ray sources in order to increase their brilliance, allowing experiments to be performed that only a few short decades ago were inconceivable. As an illustration of their impact on scientific research, 19 Nobel Prizes have been awarded to researchers working with X-rays in the fields of chemistry, physics, and medicine between 1900 and 2008.[4] X-ray absorption and emission spectroscopy are now routine experimental techniques, capable of providing unparalleled electronic and structural information on a wide variety of samples. With the continuous improvement in X-ray sources, more and more flux-demanding measurements can be performed. An example of this are time-resolved X-ray experiments where enough X-ray photons can now be collected on shorter and shorter time scales, allowing researchers to obtain detailed information on short-lived species as they evolve in time. This chapter will introduce various X-ray sources, with an emphasis on application to time-resolved measurements. This topic has been covered by several recent reviews in the literature,[5,6] so only a general overview will be presented here.

There are two approaches to performing time-resolved measurements of any kind. The first is to measure using a continuous source and a fast detector, where the time resolution is given by how quickly the detector can be operated. The second is to use a pulsed source and to change the timing between the initial modulation and the pulsed source. Both techniques have been used to perform time-resolved X-ray measurements on time scales varying from femtoseconds ($10^{-15}$ s) through to minutes or hours. This chapter will introduce how various time-resolved techniques have been used with the various X-ray sources available, including laboratory sources and third- and fourth-generation light sources. An emphasis will be placed on time-resolved X-ray spectroscopy measurements.

## 1.2   LABORATORY SOURCES

The bulk of the Nobel Prizes awarded to scientists performing X-ray related research have been performed using laboratory sources. In general, these types of sources consist of accelerating high-energy electrons into a solid or liquid sample, with the resulting emission of high-energy photons. The electron energy is converted into photon energy. This description covers both X-ray tubes, rotating anode X-ray tubes, and laser-based plasma sources.

### 1.2.1   X-RAY TUBES

The very first X-rays generated by Röntgen used a simple evacuated tube with a filament and a metal anode. By placing sufficient voltage across the device, electrons were accelerated into the metal, producing high-energy photons. This is still the basis for generating X-rays using X-ray tubes. These sources are used in a range of commercial devices, including medical X-ray scanners and airport security scanners. The X-ray spectrum emitted from them depends on both the voltage applied and the metal into which the electrons are accelerated.[7,8] In general, there are two distinct types of radiation emitted from such devices. The first is a continuous background of Bremsstrahlung radiation, which extends over a broad range of X-ray photon energies, with an upper limit generally defined by the electron acceleration voltage used. An example of this is shown in Figure 1.1a from a tungsten-based X-ray tube operated at 25 kV. The second type of radiation emitted from an X-ray tube is characteristic of the metal used as the anode, and generally consists of sharp line spectra

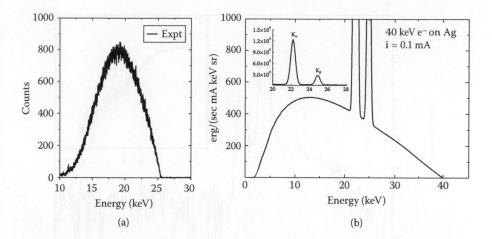

**FIGURE 1.1** Examples of X-ray tube spectra. (a) The X-ray tube spectrum from a tungsten anode operated at 25 kV and a current of 0.03 mA with Al filters to suppress lower photon energies, including the tungsten L-edges below 12 keV. The detector used was an Si(Li) solid-state detector with an energy resolution of 140 eV. (Reproduced from Mandal, A.C. et al., 2004, *Nuclear Instruments and Methods in Physics Research Section B—Beam Interactions with Materials and Atoms*, 217, 1, 104–112.[9] With permission.) (b) The emission from an Ag anode operated at 40 kV with a current of 0.1 mA showing the broad Bremsstrahlung emission extending from about 3 keV to the cutoff energy at 40 keV and the Ag Kα and Kβ emission lines at 22 and 25 keV, respectively. (Reproduced from Mandal, A.C. et al., 2002, *Nuclear Instruments and Methods in Physics Research Section B—Beam Interactions with Materials and Atoms*, 197, 3–4, 179–184.[12] With permission.)

at specific X-ray energies that are unique to the metal. An example of the emission from an X-ray tube operated at 40 kV with a silver anode is shown in Figure 1.1b. The characteristic Kα and Kβ Ag emission lines are clearly seen at 22 and 25 keV photon energies. These sources can be used for both elemental analysis, by exposing a sample to the broad-spectrum X-rays and energy-resolving the X-ray emission from the sample,[9] or for structural analysis by filtering out the broad spectrum X-rays and using the narrow emission lines to perform crystal or powder diffraction measurements. In general, X-ray tubes are capable of generating a broad range of X-ray energies, but the spectral brightness of such sources is not high, limiting their usefulness for measurements. Figure 1.2 shows the spectral brightness of some laboratory X-ray sources in comparison to the spectral brightness of synchrotron radiation sources.[10] Even with the development of rotating anode and microfocus X-ray tube sources, the spectral brightness remains several orders of magnitude below that of synchrotron-based X-ray sources. Though it is conceivable to perform time-resolved measurements with such sources, the acquisition times would be hours or days,[11] greatly limiting the time scales over which dynamics can be measured.

One approach to improving both the flux and temporal resolution of X-ray tubes is to generate the pulse of electrons using a short laser pulse incident on a photocathode

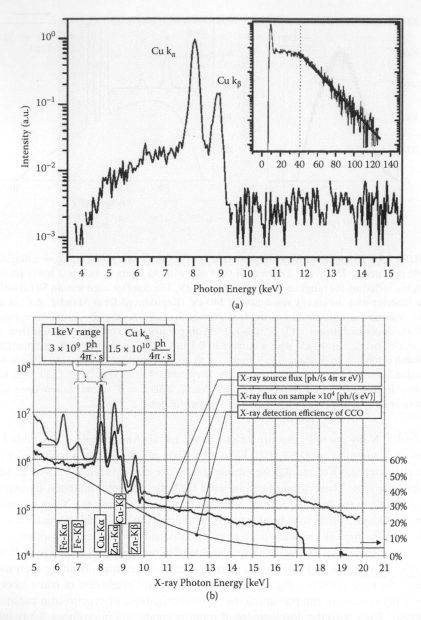

**FIGURE 1.2**   Examples of X-ray spectra generated by laser-driven plasma sources. (a) X-ray spectrum generated from copper foil showing characteristic Kα and Kβ emission lines measured with an Si-based Amptek detector. (Inset: The same spectrum measured with a scintillator + photomultiplier detector.) (Reproduced and modified from Zhavoronkov, N. et al., 2005, *Optics Letters*, 30, 13, 1737–1739.[19] With permission.) (b) X-ray spectrum measured from a brass-coated steel wire showing copper, iron, and zinc emission lines with the Bremsstrahlung tail. Included for reference are the transmission of the X-ray optic used to collect and focus the X-rays, and the detector response as a function of X-ray photon energy. (Reproduced and modified from Benesch, F. et al., 2004, *Optics Letters*, 29, 1028–1030.[11] With permission.)

and then to accelerate these electrons into the metal anode.[13] The composition of the anode determines the characteristics of the X-ray spectrum generated. This approach, called an *X-ray diode*, can generate pulses down to the nanosecond or even picosecond regime, with a repetition rate of Hz to kHz, depending on the laser source used.[14] In general, these sources operate with metal photocathodes, such as Al, which require excitation with UV light to produce photoelectrons. Since this process is resonant, a single laser photon can generate a photoelectron and high fluence is not required. Sufficient photoelectrons can be generated with a few mJ/cm$^2$.[13] The advantage to this type of X-ray diode is that it provides an inherently synchronized laser source that can be used to excite samples by splitting the laser into two beams: one for photoexcitation of the sample and one for photoelectron generation at the cathode. This allows measurements to be performed with a time resolution limited by the convolution of the X-ray and laser pulse durations. The disadvantage to the laser diode approach is that the number of electrons generated by the photocathode limits the X-ray flux, and increasing the number of electrons generated by the photocathode lengthens the pulse duration due to Coulomb interactions between the electrons. In addition, it is still difficult to focus this X-ray source to obtain increased flux density on the sample. X-ray diodes have been used primarily to perform time-resolved X-ray diffraction measurements with time resolution down to the ps regime.[15,16]

## 1.2.2 Laser Plasma Sources

Laser plasma sources are essentially a modified version of an X-ray tube, where the electric field used to generate and accelerate the electrons is from a short, high-intensity laser pulse. By focusing a high-energy, ultrashort laser pulse into either a solid or liquid target, a pulse of high-energy X-rays can be generated.[17] The necessary intensities for this technique to generate X-rays are around $10^{15}$–$10^{16}$ W/cm$^2$.[5,6,18,19] As in an X-ray tube, the X-ray spectrum is determined by the material from which the X-rays are produced. Examples of the spectra generated from such sources are shown in Figure 1.2 for a solid Cu target[19] and a solid brass-coated steel wire.[11] A liquid Hg target has also been used to generate X-rays.[20] The X-rays are generally emitted in a $4\pi$ solid angle, and need to be collected with X-ray optics[21] to focus them onto the sample. By using multilayer[22] or polycapillary optics,[23] the X-rays can be focused to spot sizes of tens of microns, substantially increasing their flux density.

Examples of time-resolved XAS using laser plasma sources have been demonstrated by taking advantage of the ability to use a crystal to disperse the broad-bandwidth X-ray spectrum after the interaction with the sample to measure the X-ray absorption spectrum. An example of such a setup is shown in Figure 1.3, where the X-rays generated from a tungsten filament were used to measure the Fe K-edge XAS spectrum on a pulse-to-pulse basis. This setup has been applied to performing time-resolved XAS measurements with 25 ps time resolution on iron oxalate, demonstrating its capabilities.[24]

The primary advantage of the laser plasma technique is that the emitted X-rays follow the time structure of the laser pulse that generates them, allowing X-ray pulses to be created with durations down to the femtosecond regime ($10^{-15}$ seconds). Though

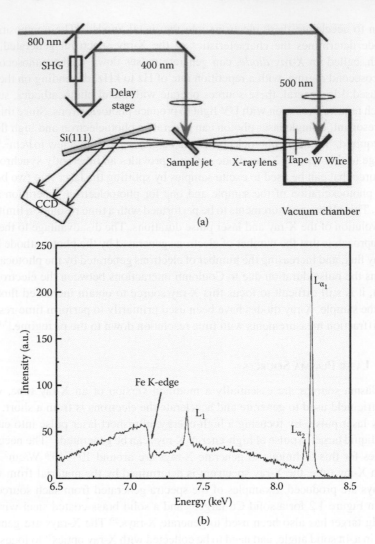

**FIGURE 1.3   (See Color Insert.)** (a) Schematic of a laser plasma X-ray source using a tungsten wire as the target, a microcapillary optic to focus the X-rays, and a dispersive detection scheme. (b) The X-ray spectrum generated by this source measured after transmission through a Fe-containing sample, showing the X-ray continuum and tungsten emission lines. (Reproduced from Chen, J. et al., 2007, *Journal of Physical Chemistry A*, 111, 38, 9326–9335.[24] With permission.)

this short pulse duration gives access to the very fastest of nuclear dynamics, the low spectral brightness and difficulty in tuning the X-ray energy makes them inefficient sources for time-resolved X-ray spectroscopy measurements and more suited to time-resolved X-ray diffraction measurements where the intense, narrow bandwidth X-ray fluorescence lines from the plasma can be used more efficiently.[22,25–28]

### 1.2.3 HIGH-HARMONIC GENERATION SOURCES

A recent technique that has received a significant amount of attention is high-harmonic generation (HHG) where an ultrashort laser pulse is focused into a gas jet, resulting in the emission of high-energy photons.[29-31] In general, this technique produces photons in the vacuum-ultraviolet (VUV) range, which covers 10–100 eV; but it has been demonstrated that it is possible to generate soft X-ray photon energies using such systems.[6,32,33] Examples of spectra generated by quasi-phase matching HHG[29] are shown in Figure 1.4, where X-ray absorption edges of Cu, Al, and Si were measured in transmission through thin foils. The measured spectra are noisy, but show a clear absorption edge characteristic of the elemental composition of the sample. These soft X-ray pulses have similar time structure to the laser pulse that creates them, which in the example shown was 20 fs in duration. Though this technique shows promise, it has remained difficult to use for time-resolved measurements both because of the low spectral brightness, but also because the X-ray photon energies generated require vacuum conditions within which to operate.

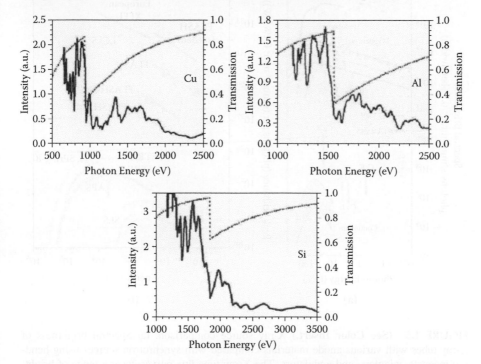

**FIGURE 1.4** X-ray spectra generated by quasi-phase matching high-harmonic generation (HHG) after transmission through several materials (solid line). The spectra show the expected decrease in transmission at the Cu $L_3$-edge (933 eV), the Al K-edge (1.56 keV), and Si K-edge (1.84 keV); the dashed lines show the calculated transmission of these samples. (Reproduced from Seres, E. et al., 2009, *Applied Physics A—Materials Science and Processing*, 96, 1, 43–50.[34] With permission.)

### 1.2.4 LASER-PLASMA ACCELERATION SOURCES

Laser plasma sources can generate high-energy electrons of up to 1 GeV energy. These electrons can then be used to drive short-period undulators, which consist of periodically alternating magnetic fields, causing the electron bunch to oscillate and emit radiation. This is identical to the approach taken in synchrotron facilities to generate high-brightness X-ray beams. To date, this approach has been used successfully to generate beams with a wavelength of 18 nm (70 eV) and a peak brilliance of $1.3 \times 10^{17}$ photons/second/mrad²/mm²/0.1% bandwidth, which is comparable in brightness to that of a synchrotron facility (see Figure 1.5).[35]

A second approach is to use the plasma field itself as the undulator replacement, which allows stronger magnetic fields to be obtained. By tilting the pulse front of the laser that generates the plasma, X-rays of 4.35 keV were obtained with the ability to

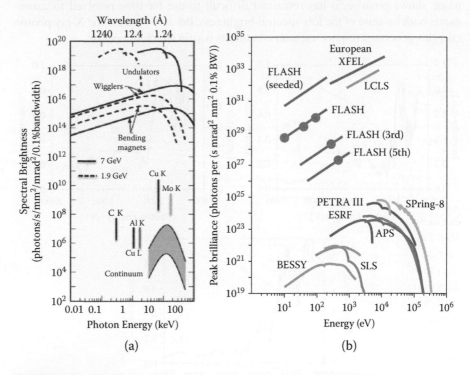

**FIGURE 1.5   (See Color Insert.)** X-ray source comparison. (a) Spectral brightness of X-ray tubes with various anode materials compared with synchrotron sources using bending magnets, wigglers, and undulators. The X-ray tube flux curves cover a range of brightness depending on the tube design. The synchrotron flux curves are shown for two different electron energies. (Plot reproduced and modified from Thompson, A.C. et al., 2009, *X-ray Data Booklet*, Lawrence Berkeley National Laboratory, Berkeley, CA.[10]) (b) Peak brilliance comparison between facility X-ray sources, ranging from 3rd-generation synchrotron sources (BESSY, SLS, SPring-8, APS, ESRF, PETRA III) and XFEL sources (FLASH), and calculated XFEL sources (LCLS, European XFEL). The blue dots are measured values from FLASH. (Reproduced from Ackermann, W. et al., 2007, *Nature Photonics*, 1, 6, 336–342.[49] With permission.)

control their polarization.[36] Though these types of sources provide promise that the goal of a tabletop X-ray free electron laser will one day be possible,[37] they are still very much in the development phase.

### 1.2.5 LABORATORY X-RAY SOURCE SUMMARY

Laser technology is increasing at a rapid pace, with higher pulse energies, shorter pulse durations, and higher photon energies continuously improving.[38,39] The current state of the art of laser-based X-ray generation is also continuously changing. The primary advantage of laser-based sources is their ability to generate extremely short duration X-ray pulses inherently synchronized to the optical laser source that generates them. These femtosecond X-ray pulses are ideal for measuring ultrafast nuclear dynamics, where the motion of nuclei can be resolved. For this type of measurement to be meaningful, an equally ultrafast excitation must take place, generally restricting excitation techniques to those feasible with ultrafast laser sources. This limitation is not as restrictive as it may seem since the spectrum of ultrafast laser sources cover the ultraviolet to the far infrared, allowing excitations to span everything from highly excited electronic states to low-frequency phonon modes in the condensed phase. Once the dynamics of interest move beyond time scales of a few picoseconds, laser-based X-ray sources begin to suffer from low average flux, reducing their advantages in comparison to accelerator-based sources, which we will introduce in the next section.

## 1.3 ACCELERATOR SOURCES

Large-scale facility accelerator-based sources of X-rays are by far the most accessible facilities for performing X-ray spectroscopic measurements. These are generally user facilities where scientists can obtain measurement time to perform their experiments. Though the initial investment can be substantial, and generally requires facilities to be built on the national or international level, once the facility is operational they attract users from a broad range of scientific disciplines. In this section we will briefly describe these facilities and what is feasible at the various categories of X-ray source.

### 1.3.1 SYNCHROTRONS

Synchrotrons provide an intense source of tunable X-rays, making them an ideal location for performing X-ray spectroscopic measurements.[40] In general, synchrotrons produce high-energy photons by passing relativistic electrons through strong magnetic fields, forcing them to change direction. This motion results in the electrons losing energy in the form of electromagnetic radiation. The magnetic fields can be applied using a variety of devices including bending magnets, superconducting bending magnets, wigglers, and undulators.[41] From a user perspective, each of them has different characteristics and results in a different X-ray source for experiments. Bending magnets generate a broad spectrum of X-rays, which can then be monochromatized using, for example, diffraction from a crystal monochromator.[42] The

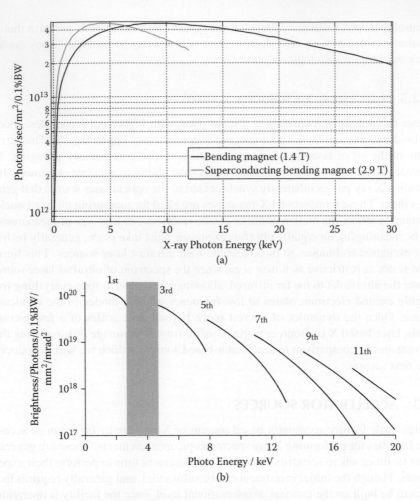

**FIGURE 1.6  (See Color Insert.)** (a) X-ray spectra generated by a bending magnet (red curve) and a superconducting bending magnet (blue curve) at the Swiss Light Source synchrotron. Flux curves calculated using XOP. (b) Calculated X-ray spectrum showing the harmonics of an in-vacuum, minigap undulator used at the microXAS beamline at the Swiss Light Source. The gray bar shows a gap in the X-ray spectrum where no X-ray photons are generated.

maximum X-ray flux in a bending magnet X-ray spectrum depends on the magnetic field strength of the magnet and the electron energy in the synchrotron ring. An example of a bending magnet spectrum at the Swiss Light Source 3rd-generation synchrotron is shown in Figure 1.6.

Bending magnets are not restricted to generating photons in the X-ray spectrum. They can also generate light in the infrared spectrum[43,44] or the vacuum-ultraviolet spectrum,[45] making them very general purpose photon sources at accelerators. If superconducting magnets are used to generate the magnetic field, the peak in the X-ray spectrum shifts to higher energies (see Figure 1.6).[46] Undulators and wigglers consist of periodic arrays of permanent magnets, which force the electrons into an oscillatory trajectory as they pass through them. The result is a collimated, spatially intense beam of X-rays. The spectrum of an undulator from the Swiss Light Source is shown in Figure 1.6. In general, undulator beam lines produce the highest spectral brilliance, with average photon fluxes exceeding $10^{12}$ photons/second on the sample (see Figure 1.5). These so-called insertion devices also generate higher-order harmonic photons, which allow them to be used at higher photon energies, albeit with a loss in X-ray flux. At an X-ray spectroscopy beam line, the X-ray photons are generally monochromatized using a crystal monochromator to a $\Delta E/E$ bandwidth of 0.01–0.03%. This energy resolution is on the order of an eV in the hard X-ray regime, which corresponds to the core hole lifetime broadening contribution in an X-ray spectrum, making it the ideal measurement bandwidth. It is possible to use higher diffraction orders from the monochromator crystals, allowing higher-energy resolution measurements to be obtained for techniques such as resonant inelastic X-ray scattering (RIXS)[47] or inelastic X-ray scattering (IXS).[48]

The monochromatic X-rays can then be focused onto the sample using reflective achromatic X-ray optics, such as Kirkpatrick–Baez mirrors.[50,51] Another approach is to maintain the broad bandwidth incident beam ($\Delta E/E\sim 1\%$) and to disperse the X-rays after interaction with the sample.[52,53] In all cases, the goal is to measure the X-ray absorption or emission spectrum from the sample by monitoring either the transmitted X-rays through the sample, or the X-rays emitted/scattered from the sample. See Chapter 4 for further details.

The fill pattern of the electrons within the synchrotron storage ring determines the time structure of the X-rays. In general, this consists of electron bunches separated by a few nanoseconds, with an entire revolution of the ring taking hundreds of nanoseconds to microseconds (see Figure 1.7) depending on the size of the storage ring. When a slow detector is used, the X-ray flux appears as a continuous stream of photons. By reading out the detector on time scales of milliseconds or seconds, the evolution of a sample can be recorded. If a faster detector is used the individual X-ray pulses can be observed, which allows the experiment to monitor sample evolution on much faster time scales. If the X-ray pulses are to be used in a pump-probe experiment, where the sample is initially excited using an ultrafast laser pulse and then subsequently probed after some adjustable time delay, it is necessary to be able to isolate the X-rays from a single X-ray pulse. This can either be done using very fast detectors with sub-ns time resolution, or by taking advantage of the ability of synchrotrons to control the electron fill pattern within the storage ring. An example of such an approach is shown in Figure 1.7 where the Swiss Light Source uses a 500 MHz radio-frequency source as the fundamental structure for its fill pattern. Bunches 1–400 are filled with 1 mA of electrons, and bunches 401–480 are left unfilled as an ion-clearing gap. An isolated bunch, sometimes called the *camshaft*, can be placed into this gap. This allows detectors with a time resolution of 10–20 ns to isolate the X-rays from the isolated bunch. The result is that measurements can then be performed where the time resolution is limited to the pulse duration of the isolated bunch, which is usually around 100 ps.[54–56] This means that by judicious choice of detectors, X-ray spectroscopy measurements at synchrotrons are capable of covering an extremely broad range of time scales, from picoseconds out to seconds and beyond.[57] By taking advantage of the ability to microfocus the X-rays to

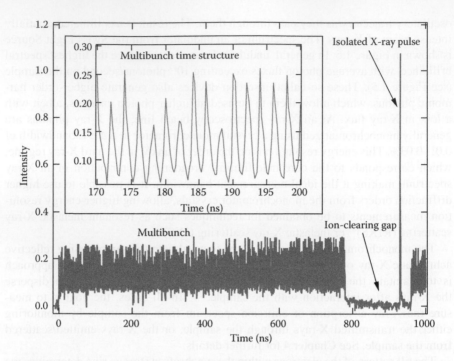

**FIGURE 1.7** **(See Color Insert.)** X-ray pulse structure as a function of time at the Swiss Light Source. The multibunch structure consists of pulses separated by 2 ns, followed by an ion-clearing gap in which a single pulse can be placed. The structure repeats at an interval of 960 ns, giving a repetition rate for the isolated pulse of 1.04 MHz.

very small spot sizes (10–20 µm) and the availability of high-power (10–20 W), high-repetition rate (MHz) lasers, pump-probe X-ray absorption and emission measurements at third-generation light sources have seen a recent increase in popularity thanks to their ability to obtain high signal-to-noise measurements on photoexcited samples.[53,58–61] The primary restriction to performing these types of pump-probe measurements is the limited time resolution that can be achieved. In the following sections we will discuss several methods of improving the time resolution of such measurements.

### 1.3.2 LOW-ALPHA MODE AT SYNCHROTRONS

One method of obtaining shorter X-ray probe pulses from synchrotron facilities is to operate them in a so-called low-alpha mode.[62–64] This is a special operation mode of the synchrotron where the bunch current is usually reduced, often by orders of magnitude, and negative dispersion in the electron beam is introduced, resulting in the generation of shorter X-ray pulses. The pulses are often shorter by a factor of 2–5 over normal operation. Several facilities offer this mode regularly throughout the year,[65] though due to the lower overall X-ray flux, the experiments that can be performed using low-alpha mode are generally limited, and the time-resolution improvement still restricts measurements to the tens of picoseconds time scale.

### 1.3.3 FEMTO-SLICING AT SYNCHROTRONS

Zholents and Zolotorev first proposed femto-slicing in 1996.[66] The idea was to overlap an intense femtosecond laser pulse with an electron bunch within a specially designed modulator in a synchrotron storage ring. This modulator should consist of a periodic array of alternating permanent magnets, as in a wiggler or undulator, where the magnetic field is designed to optimize the interaction between the laser electric field and the oscillating electrons.[67,68] The result would then be that a slice of the electron bunch has more energy, while another part has less energy. This slice would have the time scale of the femtosecond laser pulse, generally 50–100 fs. Then by passing the electrons through a magnetic chicane, the portions of the bunch with more or less energy would be spatially separated from the main bunch. The electrons then pass through an undulator, generating X-rays, and by careful spatial alignment the sliced femtosecond X-rays can be used to perform experiments. This technique was first demonstrated at the Advanced Light Source (Berkeley, California) where a bending magnet was used to generate the X-rays,[20,69] and similar systems based on undulator X-ray generation were subsequently installed at the BESSY II synchrotron[70] (Berlin, Germany) and the Swiss Light Source synchrotron[71] (Villigen, Switzerland). The X-ray spectral characteristics of these femto-slicing beam lines mimic those of the original source. The pulse duration of these sources is generally 100–200 fs, due to a certain amount of dispersion in the electron bunch between the modulator and undulator devices. The repetition rate of the source is entirely dependent on the laser system used to perform the slicing, generally 1–5 kHz. The primary drawback to performing pump-probe X-ray spectroscopy measurements with these sources is the low per-pulse monochromatic flux available.[72] Due to the fact that the slicing interaction selects out a 100 fs slice from a 100 ps electron bunch, three orders of magnitude of photons are immediately lost. This, combined with the low repetition rate of the slicing laser sources, means that the average X-ray flux at such sources is $10^6$–$10^7$ photons/second/1% bandwidth, and around $10^4$ photons/second/0.015% bandwidth. Time-resolved XAS measurements have been successfully performed with femto-slicing sources, but they generally require long acquisition times and significant signal levels to make the experiments feasible.[73–76] The only reason such experiments are possible in the first place is due to the excellent stability of the femto-slicing sources.[71] Because these sources generally excite the sample with a portion of the laser that performs the laser-electron slicing process, there is little to no timing jitter to reduce the time resolution of the measurements.

### 1.3.4 X-RAY FREE ELECTRON LASERS

The recent development of hard X-ray free electron lasers (XFELs) has revolutionized the field of ultrafast time-resolved X-ray measurements.[77–79] These sources consist of a high-energy (GeV) electron bunch injected into a series of undulators that are hundreds to thousands of meters long. The oscillation of the electrons in the initial part of the undulators causes radiation to be emitted, as in a synchrotron. As the radiation and electrons copropagate, the radiation field builds up and the electrons start to interact with the radiation, causing a microbunch structure to appear with the wavelength of the radiation (see Figure 1.8).

This microbunch structure radiates coherent X-ray photons, the intensity of which build up exponentially in a process called *self-amplified spontaneous emission* (SASE). The result is an intense, spatially coherent beam of femtosecond X-ray pulses, which can be used for experiments. The FLASH VUV-soft X-ray free electron laser located at DESY in Hamburg has been the prototype for this kind of facility since it started user operation in mid-2005.[40,81,82] It was soon followed by the Linac Coherent Light Source (LCLS) (Stanford, California), which was the first hard X-ray free electron laser in operation.[83] LCLS operates both in the soft X-ray and hard X-ray regimes, with experimental stations dedicated to various fields of research.[84] Recently, the SACLA XFEL facility at SPring-8 in Japan and the FERMI@Elettra facility in Italy, also began operation,[85,86] and several XFEL projects are underway worldwide, including machines in Germany,[87] Korea,[35,88,89] and Switzerland.[36,90] Here we will present a brief overview of the distinctive characteristics of the radiation generated by the SASE process as compared to a synchrotron facility.

Because of the spontaneous nature of the SASE process, the radiation generated by XFELs has a large variance in pulse energy (photon flux), photon energy (spectrum), and pulse arrival time. The result is large pulse-to-pulse fluctuations in these parameters. When a monochromator is inserted in the beam to perform spectroscopic measurements, the photon energy instability adds to the X-ray flux instabilities at the sample position.[91] The most straightforward approach to solving this problem has been to measure the incident X-ray flux as accurately as possible, allowing these fluctuations to be normalized out.[92,93] The drawback to this approach is that these intensity monitors need to be linear over many orders of magnitude of photon flux, which is difficult to achieve.[91] A second approach that has been taken is to try to acquire as much data as possible in a single measurement, using techniques such as dispersive X-ray emission[94,95] and absorption.[53,96] Though these approaches help ameliorate the problems, they introduce further difficulties, such as measuring an accurate incident photon energy spectrum.[96,97] The most promising approach to reducing the instability of the XFEL beam at the sample position is to seed the free electron laser, preferentially selecting a portion of the photon spectrum to initiate the lasing process.[98] The result of this is a significant improvement in the spectral stability of the XFEL beam, at a cost of a factor of approximately 5–10 in photon flux.[80] The effect this technique has on the X-ray spectrum of the XFEL is shown in Figure 1.8. Though the seeded X-ray spectrum is not properly monochromatic, as it has a tail that extends a few eV to lower photon energies, it greatly enhances the pulse-to-pulse energy stability of the photon beam through a monochromator. It should be emphasized here that XFEL facilities are under continuous improvement, and the operational characteristics vary from facility to facility.

The primary feature of an XFEL is the enormous number of photons ($10^{11}$–$10^{12}$ photons/pulse) in a pulse with durations of 10–100 fs, or even sub-fs.[99] This makes their peak X-ray brilliance unparalleled (see Figure 1.5). Both SACLA and LCLS operate at low repetition rates of 100–120 Hz, which makes their average X-ray flux similar to that of 3rd-generation synchrotron facilities. The experiments that have benefited the most from the development of XFEL facilities have been those that

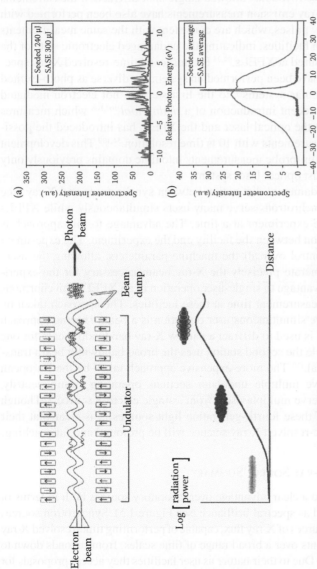

**FIGURE 1.8 (See Color Insert.)** Left: The formation of the microbunch structure in the electron pulse as it propagates through the undulators in an XFEL. (Reproduced from Sonntag, B., 2001, *Nuclear Instruments and Methods in Physics Research Section A—Accelerators Spectrometers Detectors and Associated Equipment*, 467–468, 8–15.[79] With permission.) Right: (a) Single-pulse spectra emitted from the LCLS in both unseeded (red curve) and self-seeded (blue curve) modes. Note the significant spectral narrowing when operating in the self-seeded mode. (b) Average spectra for LCLS operating in unseeded (red curve) and self-seeded (blue curve) modes. (Reproduced from Amann, J. et al., 2012, *Nature Photonics*, 6, 10, 693–698.[80] With permission.)

either take advantage of the large number of photons per pulse to perform single-shot or nonlinear X-ray experiments, or those that take advantage of the ultrashort pulse durations to perform measurements on the fs time scale. Taking advantage of the ability of the ultrashort X-ray pulse to diffract from the crystalline structure of the sample before it is destroyed has allowed single-shot diffraction measurements to be performed.[100–104] X-ray emission measurements have also been performed with high-peak intensity X-ray pulses, which are consistent with the same measurements performed at synchrotron facilities, indicating the undamaged electronic state of the sample may also be measured at XFELs.[94,95] Femtosecond time-resolved X-ray spectroscopy measurements have been performed on systems as diverse as photo-excited spin transitions in molecular systems[91] to the first steps in hot electron-mediated photocatalysis.[105,106] The recent introduction of a *timing tool*,[107,108] which measures the timing jitter between the optical laser and the XFEL, has introduced the possibility of performing measurements with 10 fs time resolution.[54,109] This development has catapulted XFEL pump-probe measurements into time domains previously only accessible with an ultrafast optical laser system.

Perhaps the most fundamental distinction between synchrotrons and X-ray free electron lasers is that synchrotrons serve many users simultaneously while XFELs are dedicated to a single experiment at a time. The advantage to this approach is unprecedented cooperation between the facility and the experiment. The experiment now has fine-grained control over all the machine parameters, allowing the users to tune the XFEL to generate precisely the X-ray beam necessary for the experiment. The obvious disadvantage to single-user operation of an XFEL is an enormous reduction in available measurement time at these facilities. The approach taken by the LCLS facility to have simultaneous user operation is a beam-sharing approach where a diamond crystal is used to diffract a narrow X-ray bandwidth beam for one experimental station while the second station uses the broad bandwidth beam transmitted through the crystal.[110] The more expensive approach taken by the European XFEL facility is to have multiple undulator sections operating simultaneously, allowing the facility to serve multiple users from a single electron source.[87] Though we are in the infancy of these fourth-generation light sources, it is clear that their effect on the field of time-resolved X-ray science will be profound and far-reaching.

### 1.3.5 ACCELERATOR X-RAY SOURCE SUMMARY

Accelerator facilities have a clear advantage over laboratory sources both in terms of spectral tunability as well as spectral brilliance (see Figure 1.5). Synchrotron sources represent stable, bright sources of X-ray flux, capable of performing time-resolved X-ray spectroscopy measurements over a broad range of time scales, from seconds down to hundreds of picoseconds. Due to their nature as user facilities they accept proposals for measurement time from scientists at regular intervals, which makes them accessible to users worldwide. The introduction of X-ray free electron lasers has enabled X-ray measurements to move into the femtosecond time domain with an enormous increase of six to seven orders of magnitude in the X-ray flux per pulse over 3rd-generation synchrotron sources. Though it is currently technically challenging to perform measurements at XFELs, the rapid pace of advancement in experimental techniques at these facilities is breathtaking. XFELs are similar to synchrotrons in that they are user facilities, but due to their single-user mode of operation, the number of measurement hours available is extremely limited, making the competition for measurement time very intense. The introduction of beam-sharing techniques[110] and future XFEL sources[87,89,90] will somewhat alleviate this problem, but not for several years to come.

## 1.4 X-RAY DETECTORS

As mentioned in the Introduction there are two approaches to performing time-resolved X-ray measurements: the first is to use a pulsed source of X-rays to perform pump-probe measurements, while the second is to use a continuous flux of X-rays with a fast detector. In this section we will briefly discuss some specific X-ray detectors that have been used to perform time-resolved measurements with specific X-ray spectroscopy techniques.

### 1.4.1 POINT DETECTORS

This detector description covers a broad array of general-purpose detectors that have been used for time-resolved X-ray measurements including diodes,[111,112] avalanche photodiodes (APDs),[113,114] and scintillators.[115] They generally function by measuring a signal proportional to the X-ray flux, though care must be taken to ensure they are measuring the flux linearly. In general, these devices are used with amplifiers, which also must be checked for linearity. These are the simplest devices for measuring X-ray flux and have been used for decades for this purpose. The ability to use them in a time-resolved fashion depends on the response time of the detector+amplifier, and is generally possible down into the nanosecond regime. This allows them to perform measurements at synchrotrons with hybrid fill patterns (see Figure 1.7) where a single pulse is isolated from the remaining X-ray signal by several tens of nanoseconds. It is also possible to operate these types of detectors with more sensitive amplifiers, reducing their time resolution but allowing them to acquire X-ray signals with millisecond to second time resolution, which is often fast enough to resolve reaction intermediates from chemical kinetics measurements.[57,116,117] Point detectors also work efficiently with Johann-geometry[118] X-ray emission spectrometers where an analyzer crystal diffracts and focuses a narrow X-ray energy bandwidth from X-rays scattered or emitted from the sample.[119,120]

An alternate approach is to scan the incident X-ray photon energy extremely quickly, for example, covering a 1 keV range in 1 second.[121–125] For this technique to work the detectors used must also be capable of being read out on a millisecond or better time scale, allowing the experiment to read the X-ray flux signal and the photon energy simultaneously, without any attempt to synchronize the measurement.

### 1.4.2 TWO-DIMENSIONAL AND STRIP DETECTORS

Strip and 2D detectors provide spatial details on X-ray signals that are not available from point detectors. In general, for X-ray spectroscopy this is not necessary, but for certain techniques it is indispensable. Both X-ray absorption[52,53,96] and X-ray emission[126,127] can be performed as dispersive measurements, where the X-ray photon spectrum is spatially dispersed using a crystal or multilayer. This requires either a strip or 2D detector to resolve the spatially dispersed X-ray spectrum. Charge-coupled detectors (CCDs) have been used for decades at synchrotron sources and are a very mature technology but are not generally capable of being used for measurements below 1 s time resolution. The ability to gate strip or 2D detectors, allowing them to measure X-rays within a narrow time window (for example, from only the isolated bunch in a hybrid fill pattern; see Figure 1.7), is a recent advancement, and is generally still limited to specific types of detectors.[128–134] This ability has allowed pixel detectors to be used as detectors for pump-probe X-ray spectroscopy measurements with 100 ps time resolution.[135] They have also been used as detectors for the monitoring of chemical kinetics, where dispersive measurement of either high-energy resolution off-resonant spectroscopy (HEROS)[136,137] can monitor the unoccupied electronic states or resonant X-ray emission spectroscopy (RXES)[119,138] can monitor the unoccupied and occupied electronic states as the chemical reaction evolves in time.[139] These techniques are described in more detail in Chapter 3.

Measurements at XFELs are not nearly as straightforward as at synchrotrons. The significant increase in peak intensity has generally made most X-ray detectors unusable at XFELs, and those that do work are generally not linear over the many orders of magnitude that the X-ray flux can cover.[91] This has required development of new detector technologies.[140–142] Particular attention has been paid to 2D detectors, primarily for scattering or diffraction measurements,[143] but which have also been used for X-ray emission spectroscopy.[94,144] One particular approach that shows promise is adaptive gain switching, which has a dynamic gain of up to $10^4$ 12 keV photons, while still being capable of performing single-photon counting at low X-ray flux levels.[145,146]

### 1.4.3 X-ray Streak Cameras

One of the approaches taken to improve the time resolution of measurements at synchrotrons is to use fast detectors to resolve the time structure within an isolated X-ray pulse. This is similar to using a fast detector with a continuous probe source, where now the continuous probe source is a 100 ps duration isolated X-ray pulse. These detectors are based on streak cameras, where the X-ray pulse generates a pulse of electrons from a photocathode, and a fast voltage ramp disperses the electrons spatially on a detector such as a microchannel plate-phosphor-CCD combination or direct detection with a CCD.[6,147–149] Because of the time dependence of the voltage ramp, this maps the spatial location of the electron signal on the 2D detector to time. This technique has been used effectively in the soft X-ray energy range to perform time-resolved XAS measurements on ultrafast melting of carbon and silicon.[150–151] Due to the lower efficiency of photoelectron generation with hard X-rays, streak cameras have generally been limited to use with X-ray photons below 2 keV.

## 1.5 SUMMARY

The ability to perform time-resolved X-ray spectroscopic measurements enables experiments to apply a broad range of techniques to a variety of dynamical systems.

As the forthcoming chapters will demonstrate, the ability to resolve the evolution in time of a system provides much greater information than a steady-state measurement can provide. The time scale on which a system is changing determines the X-ray source and detector appropriate for its measurement. As the previous sections have made clear, there are large varieties of X-ray sources available that cover a broad range of time scales, and there are an equivalently large number of X-ray detectors that can be applied to a measurement. Not all X-ray sources are ideal for all measurements, and not all X-ray detectors are ideal for all X-ray sources. It is hoped that the previous sections will enable scientists to properly choose the most appropriate combinations for their time-resolved X-ray measurements.

## REFERENCES

1. Röntgen, W.C. On a new kind of rays. *Science*. 3, 59 (1896) 227–231.
2. Röntgen, W.C. On a new kind of rays. *Nature*. 53, 1369 (1896) 274–276.
3. Lytle, F.W. The EXAFS family tree: A personal history of the development of extended X-ray absorption fine structure. *Journal of Synchrotron Radiation*. 6, 3 (1999) 123–134.
4. Robinson, A.L. and Plummer, B. eds. 2008. *Science and Technology of Future Light Sources*. U.S. Department of Energy.
5. Bressler, C. and Chergui, M. Ultrafast X-ray absorption spectroscopy. *Chemical Reviews*. 104, (2004) 1781–1812.
6. Pfeifer, T. et al. Femtosecond X-ray science. *Reports on Progress in Physics*. 69, 2 (2006) 443–505.
7. Kaye, G. The composition of the X-rays from various metals. *Proceedings of the Royal Society of London. Series A*. 93, 653 (1917) 427–442.
8. Ulrey, C.T. An experimental investigation of the energy in the continuous X-ray spectra of certain elements. *Physical Review*. 11, 5 (1918) 401–410.
9. Mandal, A.C. et al. Bremsstrahlung excited standardless EDXRF analysis. *Nuclear Instruments and Methods in Physics Research Section B—Beam Interactions with Materials and Atoms*. 217, 1 (2004) 104–112.
10. Thompson, A.C. et al. 2009. *X-ray Data Booklet*. Lawrence Berkeley National Laboratory, Berkeley, CA.
11. Benesch, F. et al. Ultrafast laser-driven X-ray spectrometer for X-ray absorption spectroscopy of transition metal complexes. *Optics Letters*. 29, (2004) 1028–1030.
12. Mandal, A.C. et al. Self-absorption correction factor for a sample excited by the bremsstrahlung radiation. *Nuclear Instruments and Methods in Physics Research Section B—Beam Interactions with Materials and Atoms*. 197, 3–4 (2002) 179–184.
13. Tomov, I.V. et al. Nanosecond hard X-ray source for time resolved X-ray diffraction studies. *Review of Scientific Instruments*. 66, 11 (1995) 5214–5217.
14. Tomov, I.V. et al. Ultrafast time-resolved transient structures of solids and liquids studied by means of X-ray diffraction and EXAFS. *Journal of Physical Chemistry B*. 103, 34 (1999) 7081–7091.
15. Tomov, I. and Rentzepis, P. Ultrafast time-resolved transient structures of solids and liquids by means of extended X-ray absorption fine structure. *Chem. Phys. Chem*. 5, 1 (2004) 27–35.
16. Tomov, I.V. and Rentzepis, P.M. Ultrafast X-ray determination of transient structures in solids and liquids. *Chemical Physics*. 299, 2–3 (2004) 203–213.
17. Rousse, A. et al. Efficient K X-ray source from femtosecond laser-produced plasmas. *Physical Review E*. 50, 3 (1994) 2200–2207.

18. Zamponi, F. et al. Femtosecond hard X-ray plasma sources with a kilohertz repetition rate. *Applied Physics A—Materials Science and Processing*. 96, 1 (2009) 51–58.
19. Zhavoronkov, N. et al. Microfocus Cu K-alpha source for femtosecond X-ray science. *Optics Letters*. 30, 13 (2005) 1737–1739.
20. Reich, C. et al. Ultrafast X-ray pulses emitted from a liquid mercury laser target. *Optics Letters*. 32, (2007) 427–429.
21. Bargheer, M. et al. Comparison of focusing optics for femtosecond X-ray diffraction. *Applied Physics B—Lasers and Optics*. 80, 6 (2005) 715–719.
22. Zamponi, F. et al. Femtosecond powder diffraction with a laser-driven hard X-ray source. *Optics Express*. 18, 2 (2010) 947–961.
23. Tomov, I.V. et al. Efficient focusing of hard X-rays generated by femtosecond laser driven plasma. *Chemical Physics Letters*. 389, 4–6 (2004) 363–366.
24. Chen, J. et al. Transient structures and kinetics of the ferrioxalate redox reaction studied by time-resolved EXAFS, optical spectroscopy, and DFT. *Journal of Physical Chemistry A*. 111, 38 (2007) 9326–9335.
25. Bargheer, M. et al. Recent progress in ultrafast X-ray diffraction. *Chem. Phys. Chem*. 7, 4 (2006) 783–792.
26. Linde, von der, D. et al. "Ultrafast" extended to X-rays: Femtosecond time-resolved X-ray diffraction. *Zeitschrift fur Physikalische Chemie*. 215, 12 (2001) 1527–1541.
27. Sokolowski-Tinten, K. et al. Femtosecond X-ray measurement of coherent lattice vibrations near the Lindemann stability limit. *Nature*. 422, 6929 (2003) 287–289.
28. Sokolowski-Tinten, K. and Linde, von der, D. Ultrafast phase transitions and lattice dynamics probed using laser-produced X-ray pulses. *Journal of Physics-Condensed Matter*. 16, 49 (2004) R1517–R1536.
29. Corkum, P.B. Plasma perspective on strong field multiphoton ionization. *Physical Review Letters*. 71, 13 (1993) 1994–1997.
30. Corkum, P.B. and Krausz, F. Attosecond science. *Nature Physics*. 3, 6 (2007) 381–387.
31. Drescher, M. et al. X-ray pulses approaching the attosecond frontier. *Science*. 291, 5510 (2001) 1923–1927.
32. Seres, E. and Spielmann, C. Ultrafast soft X-ray absorption spectroscopy with sub-20-fs resolution. *Applied Physics Letters*. 91, 12 (2007) 121919.
33. Seres, E. et al. Generation of coherent soft-X-ray radiation extending far beyond the titanium L edge. *Physical Review Letters*. 92, 16 (2004) 163002–1.
34. Seres, E. et al. Time resolved spectroscopy with femtosecond soft-X-ray pulses. *Applied Physics A—Materials Science and Processing*. 96, 1 (2009) 43–50.
35. Fuchs, M. et al. Laser-driven soft-X-ray undulator source. *Nature Physics*. 5, 11 (2009) 826–829.
36. Schnell, M. et al. Optical control of hard X-ray polarization by electron injection in a laser wakefield accelerator. *Nature Communications*. 4 (2013) 2421-1–2421-6.
37. Grüner, F. et al. Design considerations for table-top, laser-based VUV and X-ray free electron lasers. *Applied Physics B-Lasers and Optics*. 86, 3 (2007) 431–435.
38. Backus, S. et al. High power ultrafast lasers. *Review of Scientific Instruments*. 69, (1998), 1207–1223.
39. Keller, U. Recent developments in compact ultrafast lasers. *Nature*. 424, 6950 (2003), 831–838.
40. Margaritondo, G. A primer in synchrotron-radiation. Everything you wanted to know about SEX (Synchrotron Emission X-rays) but were afraid to ask. *Journal of Synchrotron Radiation*. 2, (1995) 148–154.
41. Wilmott, P. 2011. *An Introduction to Synchrotron Radiation: Techniques and Applications*. West Sussex: Wiley.

42. Ishikawa, T. et al. High-resolution X-ray monochromators. *Nuclear Instruments and Methods in Physics Research Section A—Accelerators Spectrometers Detectors and Associated Equipment.* 547, 1 (2005) 42–49.
43. Carroll, L. et al. Ultra-broadband infrared pump-probe spectroscopy using synchrotron radiation and a tuneable pump. *Review of Scientific Instruments.* 82, 6 (2011) 063101.
44. Lerch, P. et al. IR beamline at the Swiss Light Source. (2012) 012003-1–012003-8.
45. Johnson, M. et al. Vacuum ultraviolet beamline at the Swiss Light Source for chemical dynamics studies. *Nuclear Instruments & Methods in Physics Research Section A-Accelerators Spectrometers Detectors and Associated Equipment.* 610, 2 (2009) 597–603.
46. Gabard, A., George, D., Negrazus, M., Rivkin, L., Vrankovic, V., and Kolokolnikov, Y. A 2.9 Tesla room temperature superbend magnet for the Swiss Light Source at PSI. *Proceedings of IPAC2011*, San Sebastián, Spain (2011) 3038–3040.
47. Ament, L.J.P. et al. Resonant inelastic X-ray scattering studies of elementary excitations. *Reviews of Modern Physics.* 83, 2 (2011) 705–767.
48. Krisch, M. and Sette, F. Inelastic X-ray scattering from phonons. *Light Scattering in Solids Ix.* 108, (2007) 317–369.
49. Ackermann, W. et al. Operation of a free-electron laser from the extreme ultraviolet to the water window. *Nature Photonics.* 1, 6 (2007) 336–342.
50. Iida, A. and Hirano, K. Kirkpatrick–Baez optics for a sub-mu m synchrotron X-ray microbeam and its applications to X-ray analysis. *Nuclear Instruments & Methods in Physics Research Section B-Beam Interactions with Materials and Atoms.* 114, (1996) 149–153.
51. Kodama, R. et al. Development of an advanced Kirkpatrick-Baez microscope. *Optics Letters.* 21, 17 (1996) 1321–1323.
52. Pascarelli, S. et al. Energy-dispersive absorption spectroscopy for hard-X-ray micro-XAS applications. *Journal of Synchrotron Radiation.* 13, (2006) 351–358.
53. Yabashi, M. et al. Single-shot spectrometry for X-ray free-electron lasers. *Physical Review Letters.* 97, 8 (2006) 4.
54. Harmand, M. et al. Achieving few-femtosecond time-sorting at hard X-ray free-electron lasers. *Nature Photonics* 7 (2013) 215–218.
55. Lima, F.A. et al. A high-repetition rate scheme for synchrotron-based picosecond laser pump/X-ray probe experiments on chemical and biological systems in solution. *Review of Scientific Instruments.* 82, 6 (2011) 063111.
56. Saes, M. et al. A setup for ultrafast time-resolved X-ray absorption spectroscopy. *Review of Scientific Instruments.* 75, 1 (2004) 24–30.
57. Ferri, D. et al. Modulation excitation X-ray absorption spectroscopy to probe surface species on heterogeneous catalysts. *Topics in Catalysis.* 54 (2011), 1070–1078.
58. Haldrup, K. et al. Guest–host interactions investigated by time-resolved X-ray spectroscopies and scattering at MHz rates: Solvation dynamics and photoinduced spin transition in aqueous Fe(bipy)32. *Journal of Physical Chemistry A.* 116, 40 (2012) 9878–9887.
59. March, A.M. et al. Development of high-repetition-rate laser pump/X-ray probe methodologies for synchrotron facilities. *Review of Scientific Instruments.* 82, 7 (2011) 073110.
60. Stebel, L. et al. Time-resolved soft X-ray absorption setup using multi-bunch operation modes at synchrotrons. *Review of Scientific Instruments.* 82, 12 (2011) 123109.
61. Vanko, G. et al. Spin-state studies with XES and RIXS: From static to ultrafast. *Journal of Electron Spectroscopy and Related Phenomena.* 188 (2013) 166–171.
62. Abo-Bakr, M. et al. Brilliant, coherent far-infrared (THz) synchrotron radiation. *Physical Review Letters.* 90, 9 (2003), 094801.
63. Huang, X., Safranek J., and Corbett, J. Low alpha mode for SPEAR3. *Proceedings of PAC07*, Albuquerque, NM (2008) 1308–1310.

64. Robin, D. et al. Low alpha experiments at the ALS. *AIP Conference Proceedings*. 367 (1996) 181–190.
65. Feikes, J., Holldack, K., Kuske, P., and Wustefeld, G. Compressed electron bunches for THz generation—Operating BESSY II in a dedicated low alpha mode. *Proceedings of EPAC 2004*, Lucerne, Switzerland (2004) 2290–2292.
66. Zholents, A. and Zolotorev, M.S. Femtosecond X-ray pulses of synchrotron radiation. *Physical Review Letters*. 76, (1996), 912–915.
67. Ingold, G., Streun, A., Singh, B., Abela, R., Beaud, P., Knopp, G., Rivkin, L., Schlott, V., Schmidt, T., Sigg, H., van der Veen, J.F., Wrulich, A., and Khan, S. Sub-picosecond optical pulses at the SLS storage ring. *Proceedings of the 2001 Particle Accelerator Conference*, Chicago, IL (2001) 2656–2658.
68. Streun, A. et al. Sub-picosecond X-ray source FEMTO at SLS. *Proceedings of EPAC*, Edinburgh, Scotland (2006).
69. Schoenlein, R.W. et al. Generation of femtosecond pulses of synchrotron radiation. *Science*. 287, 5461 (2000) 2237–2240.
70. Khan, S. et al. Femtosecond undulator radiation from sliced electron bunches. *Physical Review Letters*. 97, 7 (2006) 074801.
71. Beaud, P. et al. Spatiotemporal stability of a femtosecond hard-X-ray undulator source studied by control of coherent optical phonons. *Physical Review Letters*. 99, 17 (2007) 174801.
72. Milne, C.J. et al. Time-resolved X-ray absorption spectroscopy: Watching atoms dance. *Journal of Physics: Conference Series*. 190 (2009) 012052.
73. Bressler, C. et al. Femtosecond XANES study of the light-induced spin crossover dynamics in an Iron(II) complex. *Science*. 323, 5913 (2009) 489–492.
74. Huse, N. et al. Femtosecond soft X-ray spectroscopy of solvated transition-metal complexes: Deciphering the interplay of electronic and structural dynamics. *Journal of Physical Chemistry Letters*. 2 (2011) 880–884.
75. Huse, N. et al. Probing the hydrogen-bond network of water via time-resolved soft X-ray spectroscopy. *Physical Chemistry Chemical Physics*. 11, 20 (2009) 3951–3957.
76. Pham, V.-T. et al. Probing the transition from hydrophilic to hydrophobic solvation with atomic scale resolution. *Journal of the American Chemical Society*. 133, 32 (2011) 12740–12748.
77. Barletta, W.A. et al. Free electron lasers: Present status and future challenges. *Nuclear Instruments and Methods in Physics Research Section A—Accelerators Spectrometers Detectors and Associated Equipment*. 618, 1-3 (2010) 69–96.
78. Pellegrini, C. The next generation of X-ray sources. *Reviews of Accelerator Science and Technology*. 03, 01 (2010) 185–202.
79. Sonntag, B. VUV and X-ray free-electron lasers. *Nuclear Instruments and Methods in Physics Research Section A—Accelerators Spectrometers Detectors and Associated Equipment*. 467–468, (2001) 8–15.
80. Amann, J. et al. Demonstration of self-seeding in a hard-X-ray free-electron laser. *Nature Photonics*. 6, 10 (2012) 693–698.
81. Bostedt, C. et al. Experiments at FLASH. *Nuclear Instruments and Methods in Physics Research Section A—Accelerators Spectrometers Detectors and Associated Equipment*. 601, 1–2 (2009) 108–122.
82. Tiedtke, K. et al. The soft X-ray free-electron laser FLASH at DESY: Beamlines, diagnostics, and end-stations. *New Journal of Physics*. 11 (2009) 023029.
83. Emma, P. et al. First lasing and operation of an angstrom-wavelength free-electron laser. *Nature Photonics*. 4, 9 (2010) 641–647.
84. *Linac Coherent Light Source*. http://lcls.slac.stanford.edu.
85. Ishikawa, T. et al. A compact X-ray free-electron laser emitting in the sub-angstrom region. *Nature Photonics*. 6 (2012) 540–544.

86. Pile, D. X-rays: First light from SACLA. *Nature Photonics*. 5, 8 (2011) 456–457.
87. *European XFEL Project*. http://www.xfel.eu/.
88. Han, J.H., Kang, H.S., and Ko, I.S. *Status of the PAL-XFEL Project*. *Proceedings of IPAC2012*, New Orleans, LA (2012) 1735–1737.
89. *PAL XFEL Project*. http://pal.postech.ac.kr/paleng/.
90. *SwissFEL*. http://www.swissfel.ch/.
91. Lemke, H.T. et al. Femtosecond X-ray absorption spectroscopy at a hard X-ray free electron laser: Application to spin crossover dynamics. *Journal of Physical Chemistry A*. 117, 4 (2013) 735–740.
92. Feng, Y. et al. A single-shot intensity-position monitor for hard X-ray FEL sources. *Proceedings of SPIE: The International Society for Optical Engineering*. 8140 (2011) 81400Q-1–81400Q-6.
93. Tono, K. et al. Single-shot beam-position monitor for X-ray free electron laser. *Review of Scientific Instruments*. 82, (2011) 023108.
94. Alonso-Mori, R. et al. 2012. Energy-dispersive X-ray emission spectroscopy using an X-ray free-electron laser in a shot-by-shot mode. *Proceedings of the National Academy of Sciences of USA*. doi:10.1073/pnas.1211384109.
95. Kern, J. et al. Simultaneous femtosecond X-ray spectroscopy and diffraction of Photosystem II at room temperature. *Science*. 340, 6131 (2013) 491–495.
96. Katayama, T. et al. Femtosecond X-ray absorption spectroscopy with hard X-ray free electron laser. *Applied Physics Letters*. 103, 13 (2013) 131105.
97. Zhu, D. et al. A single-shot transmissive spectrometer for hard X-ray free electron lasers. *Applied Physics Letters*. 101, 3 (2012) 034103.
98. Geloni, G. et al. A novel self-seeding scheme for hard X-ray FELs. *Journal of Modern Optics*. 58, 16 (2011) 1391–1403.
99. Wacker, V. et al. 2012. Sub-femtosecond hard X-ray pulse from very low charge beam at LCLS. *Proceedings of FEL 2012*, Nara, Japan. 606–609.
100. Aquila, A. et al. Time-resolved protein nanocrystallography using an X-ray free-electron laser. *Optics Express*. 20, 3 (2012) 2706–2716.
101. Barty, A. et al. Self-terminating diffraction gates femtosecond X-ray nanocrystallography measurements. *Nature Photonics*. 6, 1 (2011) 35–40.
102. Barty, A. et al. Ultrafast single-shot diffraction imaging of nanoscale dynamics. *Nature Photonics*. 2, 7 (2008) 415–419.
103. Chapman, H.N. et al. Femtosecond X-ray protein nanocrystallography. *Nature*. 469, 7332 (2011) 73–77.
104. Seibert, M.M. et al. Single mimivirus particles intercepted and imaged with an X-ray laser. *Nature*. 469, 7332 (2011) 78–81.
105. Beye, M. et al. Selective ultrafast probing of transient hot chemisorbed and precursor states of CO on Ru(0001). *Physical Review Letters*. 110, 18 (2013) 186101.
106. Dell'Angela, M. et al. Real-time observation of surface bond breaking with an X-ray laser. *Science*. 339, 6125 (2013) 1302–1305.
107. Beye, M. et al. X-ray pulse preserving single-shot optical cross-correlation method for improved experimental temporal resolution. *Applied Physics Letters*. 100 (2012) 121108.
108. Bionta, M.R. et al. Spectral encoding of X-ray/optical relative delay. *Optics Express*. 19, 22 (2011) 21855–21865.
109. Lemke, H.T. et al. Femtosecond optical/hard X-ray timing diagnostics at an FEL: Implementation and performance. (2013) 87780S–87780S-4.
110. Feng, Y. et al. Ultra-thin Bragg crystals for LCLS beam-sharing operation. *Proceedings of SPIE: International Society for Optics and Photonics*. 8778 (2013) 87780S-1–87780S-4.

111. Owen, R.L. et al. Determination of X-ray flux using silicon pin diodes. *Journal of Synchrotron Radiation.* 16 (2009) 143–151.
112. Scholze, F. et al. Mean energy required to produce an electron-hole pair in silicon for photons of energies between 50 and 1500 eV. *Journal of Applied Physics.* 84, 5 (1998) 2926–2939.
113. Baron, A.Q.R. et al. A fast, convenient, X-ray detector. *Nuclear Instruments and Methods in Physics Research Section A—Accelerators Spectrometers Detectors and Associated Equipment.* 400, 1 (1997) 124–132.
114. Baron, A.Q.R. et al. Silicon avalanche photodiodes for direct detection of X-rays. *Journal of Synchrotron Radiation.* 13, 2 (2006) 131–142.
115. Clozza, A. et al. Apparatus for time-resolved X-ray absorption spectroscopy using flash-photolysis. *Review of Scientific Instruments.* 60, 7 (1989) 2519–2521.
116. Grunwaldt, J.D. et al. X-ray absorption spectroscopy under reaction conditions: Suitability of different reaction cells for combined catalyst characterization and time-resolved studies. *Phys. Chem. Chem. Phys.* 6 (2004) 3037–3047.
117. Wang, Q. et al. Solving the structure of reaction intermediates by time-resolved synchrotron X-ray absorption spectroscopy. *Journal of Chemical Physics.* 129, 23 (2008) 234502.
118. Johann, H.H. The production-aperture X-ray spectra with the help of concave crystals. *Zeitschrift Fur Physik.* 69, 3–4 (1931) 185–206.
119. Bergmann, U. and Glatzel, P. X-ray emission spectroscopy. *Photosynthesis Research.* 102 (2009) 255–266.
120. Glatzel, P. et al. Hard X-ray photon-in photon-out spectroscopy. *Catalysis Today.* 145, 3-4 (2009) 294–299.
121. Bornebusch, H. et al. A new approach for QEXAFS data acquisition. *Journal of Synchrotron Radiation.* 6, 3 (1999) 209–211.
122. Frahm, R. Quick scanning EXAFS: First experiments. *Nuclear Instruments and Methods in Physics Research Section A—Accelerators Spectrometers Detectors and Associated Equipment.* 270, 2–3 (1988) 578–581.
123. Frahm, R. et al. The dedicated QEXAFS facility at the SLS: Performance and scientific opportunities. *AIP Conf. Proc.* 1234 (2010) 251–255.
124. Lützenkirchen-Hecht, D. et al. Piezo-QEXAFS with fluorescence detection: Fast time-resolved investigations of dilute specimens. *Journal of Synchrotron Radiation.* 8, 1 (2001) 6–9.
125. Richwin, M. et al. Piezo-QEXAFS: Advances in time-resolved X-ray absorption spectroscopy. *Journal of Synchrotron Radiation.* 8, 2 (2001) 354–356.
126. Alonso-Mori, R. et al. A multi-crystal wavelength dispersive X-ray spectrometer. *Review of Scientific Instruments.* 83, 7 (2012) 073114.
127. Szlachetko, J. et al. A von Hamos X-ray spectrometer based on a segmented-type diffraction crystal for single-shot X-ray emission spectroscopy and time-resolved resonant inelastic X-ray scattering studies. *Review of Scientific Instruments.* 83, 10 (2012) 103105.
128. Bergamaschi, A. et al. The MYTHEN detector for X-ray powder diffraction experiments at the Swiss Light Source. *Journal of Synchrotron Radiation.* 17 (2010) 653–668.
129. Bergamaschi, A. et al. Time-over-threshold readout to enhance the high flux capabilities of single-photon-counting detectors. *Journal of Synchrotron Radiation.* 18 (2011) 923–929.
130. Broennimann, C. et al. The PILATUS 1M detector. *Journal of Synchrotron Radiation.* 13 (2006) 120–130.
131. Ejdrup, T. et al. Picosecond time-resolved laser pump/X-ray probe experiments using a gated single-photon-counting area detector. *Journal of Synchrotron Radiation.* 16 (2009) 387–390.

132. Labiche, J.-C. et al. The fast readout low noise camera as a versatile X-ray detector for time resolved dispersive extended X-ray absorption fine structure and diffraction studies of dynamic problems in materials science, chemistry, and catalysis. *Review of Scientific Instruments*. 78, 9 (2007) 091301.

133. Schmitt, B. et al. Development of single photon counting detectors at the Swiss Light Source. *Nuclear Instruments and Methods in Physics Research Section A,* 518 (2004) 436–439.

134. Schmitt, B. et al. Mythen detector system. *Nuclear Instruments and Methods in Physics Research Section A—Accelerators Spectrometers Detectors and Associated Equipment.* 501, 1 (2003) 267–272.

135. Tromp, M. et al. Energy dispersive XAFS: Characterization of electronically excited states of copper(I) complexes. *Journal of Physical Chemistry B.* 117, 24 (2013) 7381–7387.

136. Kavčič, M. et al. Hard X-ray absorption spectroscopy for pulsed sources. *Physical Review B.* 87, 7 (2013) 075106.

137. Szlachetko, J. et al. High energy resolution off-resonant spectroscopy at sub-second time resolution: $(Pt(acac)_2)$ decomposition. *Chemical Communications.* 48, 88 (2012) 10898.

138. Vanko, G. et al. Probing the 3d spin momentum with X-ray emission spectroscopy: The case of molecular-spin transitions. *Journal of Physical Chemistry B.* 110, 24 (2006) 11647–11653.

139. Szlachetko, J. et al. *In situ* hard X-ray quick RIXS to probe dynamic changes in the electronic structure of functional materials. *Journal of Electron Spectroscopy and Related Phenomena.* 188 (2013) 161–165.

140. Herrmann, S. et al. CSPAD-140k: A versatile detector for LCLS experiments. *Nuclear Inst. and Methods in Physics Research, A.* 718 (2013) 550–553.

141. Kimmel, N. et al. Calibration methods and performance evaluation for pnCCDs in experiments with FEL radiation. *SPIE Optics + Optoelectronics*, T. Tschentscher and D. Cocco, eds. 8078 (2011) 80780V–80780V–11.

142. Strüder, L. et al. Large-format, high-speed, X-ray pnCCDs combined with electron and ion imaging spectrometers in a multipurpose chamber for experiments at 4th generation light sources. *Nuclear Inst. and Methods in Physics Research, A.* 614, 3 (2010) 483–496.

143. Hart, P. et al. The CSPAD megapixel X-ray camera at LCLS. *SPIE Optical Engineering + Applications*, S. P. Moeller, M. Yabashi, and S. P. Hau-Riege, eds. 8504 (2012) 85040C–85040C–11.

144. Kern, J. et al. 2012. Room temperature femtosecond X-ray diffraction of Photosystem II microcrystals. *Proceedings of the National Academy of Sciences of USA.* doi: 10.1073/pnas.1204598109.

145. Becker, J. et al. The single photon sensitivity of the adaptive gain integrating pixel detector. *Nuclear Inst. and Methods in Physics Research, A.* 694, C (2012) 82–90.

146. Mozzanica, A. et al. The GOTTHARD charge integrating readout detector: Design and characterization. *Journal of Instrumentation.* 7 (2012) C01019.

147. Feng, J. et al. A grazing incidence X-ray streak camera for ultrafast, single-shot measurements. *Applied Physics Letters.* 96, 13 (2010) 134102.

148. Feng, J. et al. An X-ray streak camera with high spatio-temporal resolution. *Applied Physics Letters.* 91 (2007) 134102.

149. Larsson, J. et al. Ultrafast X-ray diffraction using a streak-camera detector in averaging mode. *Optics Letters.* 22 (1997) 1012–1014.

150. Johnson, S.L. et al. Bonding in liquid carbon studied by time-resolved X-ray absorption spectroscopy. *Physical Review Letters.* 94, 5 (2005) 4.

151. Johnson, S.L. et al. Properties of liquid silicon observed by time-resolved X-ray absorption spectroscopy. *Physical Review Letters.* 91 (2003) 157403.

# 2 Crystal Spectrometers

## *Jean-Claude Dousse and Joanna Hoszowska*

## CONTENTS

## 2.1 INTRODUCTION

For the detection of X-rays, energy dispersive and wavelength dispersive detectors can be used. In energy dispersive detectors such as scintillators, proportional counters, and semiconductor detectors the amplitude of the electric signal provided by the detector gives the energy of the measured photons. Energy dispersive detectors are characterized by a high linearity and a good efficiency but their energy resolution $\Delta E/E$ is limited, ranging from $\sim 10^{-1}$ for scintillators to $\sim 10^{-2}$ for semiconductor detectors. In wavelength dispersive instruments (or crystal spectrometers) the photon energy is derived from the measured Bragg angle and the known spacing constant of the crystal diffraction planes. Crystal spectrometers permit measuring X-rays with a high spectral resolution ($\Delta\lambda/\lambda = \Delta E/E \cong 10^{-4}$) and a high accuracy to the order of a few ppm. However, they are also characterized unfortunately by a rather poor efficiency, which is mainly due to the tiny solid angle under which the diffracting part of the crystal lamina is viewed by the radiation source and to a smaller extent by the limited reflectivity of the crystal.

More recently, a novel family of high-energy resolution detectors has been developed. In these calorimetric detectors, also called *low temperature detectors* (LTDs) because they have to be operated at very low temperatures (some tens of mK), the particle energy is determined from the tiny increase of the detector temperature

27

resulting from the particle absorption.[1] The temperature rise $\delta T$ is related to the energy $E$ of the absorbed particle by $\delta T = E/C_{tot}$, where $C_{tot}$ represents the total heat capacity of the absorber-thermometer system. Several types of LTDs exist which differ by the method used to measure the temperature elevation. For example, in metallic magnetic calorimeters (MMCs),[2] $\delta T$ is obtained from the change of magnetization $\delta M$ of a paramagnetic sensor placed in a magnetic field. The resulting change of the magnetic flux $\delta \Phi$ through the sensor can be then transformed into a voltage or current change employing, for example, a superconducting quantum interference device (SQUID) module. Using such an MMC detector cooled down to 35 mK and equipped with an Au absorber, an Er-doped Au sensor located in a magnetic field of about 3mT and a DC SQUID, an overall full width at half maximum (FWHM) resolution of 3.4 eV could be obtained for the $K\alpha_{1,2}$ X-ray lines of Mn ($E \cong 5.9$ keV).[3] However, although promising these calorimetric detectors present key disadvantages such as their small size (about 0.06 mm$^2$ for the detector discussed in Fleischmann et al.[3]), their limited speed (typical temperature recover times are a few ms), their relatively high cost as well as the space which is needed for the cryogenic system used to cool down the detector. MMCs and other types of LTD detectors are therefore not really adequate for time resolved high-energy resolution X-ray spectroscopy of catalysts and for this reason our discussion will be focused thereafter on crystal spectrometers.

Since the pioneering work of Bragg, many crystal spectrometers based on a variety of different setups have been developed. Spectrometers using flat crystals provide the highest resolution and precision but their luminosity is extremely weak. To increase the solid angle, which is the main reason for the poor efficiency of crystal spectrometers, the crystal lamina can be bent, cylindrically or spherically. Bending the crystal results in efficiency enhancements of the order of $10^2$–$10^3$ with, however, some losses in the resolution, due to the geometrical aberrations related to imperfections in the crystal curvature and the quasi-mosaicity induced in the crystal plate by the bending torque.[4] For both flat and curved-crystal spectrometers, Bragg-type and Laue-type instruments do exist. Most commonly used setups corresponding to the Bragg case (reflection-type crystal spectrometers) are presented in Section 2.3, those related to the Laue case (transmission-type crystal spectrometers) in Section 2.4, whereas the basic principles governing the diffraction of X-rays by crystals are outlined in Section 2.2. To conclude this chapter, the potential provided by the combination of high-energy resolution crystal spectrometers with synchrotron radiation or X-ray free electron laser (XFEL) beams for the detailed investigation of the electronic structure of catalysts is discussed in Section 2.5.

## 2.2   BASIC PRINCIPLES

Independent from their specific characteristics, crystal spectrometers are all based on the Bragg law:

$$2 \cdot d_{hk\ell} \cdot \sin(\vartheta) = n \cdot \lambda = n \cdot \frac{h \cdot c}{E}, \tag{2.1}$$

where $d_{hk\ell}$ is the spacing constant of the crystal diffraction planes defined by the Miller's indices $(hk\ell)$, $\vartheta$ the Bragg angle, that is, the angle between the diffraction plane and the incoming or diffracted X-ray, $n$ an integer number corresponding to the diffraction order, $\lambda$ and $E$ the wavelength and energy of the diffracted radiation, $h$ the Planck constant, and $c$ the velocity of light in vacuum.

There are two main diffraction geometries that are employed in crystal spectrometers. Diffraction geometry shown in Figure 2.1a corresponds to the so-called symmetric Bragg case (or symmetric reflection case), while the one shown in Figure 2.1b is usually called the *symmetric Laue case* (or symmetric transmission case). In the symmetric Bragg case, the diffraction planes are parallel to the crystal surface exposed to the incoming radiation while in the symmetric Laue case they are perpendicular to it. If the crystal is cut in such a way that the diffraction planes make a tilt angle with its front surface (reflection case) or with the normal to the surface (transmission case) the diffraction geometries are referred to as the *asymmetric Bragg case* and *asymmetric Laue case*, respectively.

In the Laue case, the photons have to travel through the crystal thickness to reach the detector. As a result, the intensity of the incident and diffracted radiation is attenuated, being partly absorbed by the crystal. This absorption grows drastically when the photon energy decreases. For instance, for a 1 mm thick quartz plate, the absorption amounts to 4% at 100 keV, 46% at 20 keV, and 99% at 10 keV. For this reason, transmission-type crystal spectrometers are usually not employed for photon energies below about 15 keV. On the other hand, due to geometrical constraints related to the spectrometer design, the minimum Bragg angles sustainable by reflection-type crystal spectrometers cannot be below about 20 degrees. For standard crystals such as quartz, silicon, and germanium with typical spacing constants $2d_{hk\ell} \approx 3\text{--}5$ Å, Bragg angles $\vartheta \leq 20$ degrees correspond to photon energies $E \leq 7\text{--}12$ keV so that Bragg-type crystal spectrometers are generally used for measuring photons below about 15 keV. In that sense, Bragg-type and Laue-type crystal spectrometers can be considered as complementary high-resolution instruments.

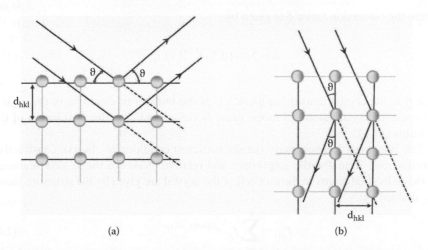

(a)                                                            (b)

**FIGURE 2.1** **(See Color Insert.)** (a) Bragg and (b) Laue diffraction geometries.

If the spacing constant $d_{hk\ell}$ of the crystal is known, the energy $E$ of the diffracted radiation can be determined from the measured Bragg angle $\vartheta$:

$$E = \frac{h \cdot c}{2 \cdot d_{hk\ell}} \cdot \frac{n}{\sin(\vartheta)}. \tag{2.2}$$

Note that the energy $E$ can be obtained directly in keV by entering in Equation (2.2) $d_{hk\ell}$ in Å and substituting for $h \cdot c$ the value of 12.398419 keV·Å deduced from the constants given in Mohr, Taylor, and Newell.[5] If the spacing constant $d_{hk\ell}$ is not accurately known, the energy $E$ of the X-ray line of interest can be determined by measuring a reference X-ray line whose energy $E_0$ can be taken from an X-ray transition energies database such as the one from Deslattes et al.[6] using the following relation deduced from Equation (2.2):

$$E = \frac{n \cdot \sin(\vartheta_0)}{n_0 \cdot \sin(\vartheta)} \cdot E_0, \tag{2.3}$$

where $n_0$ and $\vartheta_0$ stand for the diffraction order and Bragg angle of the X-ray line taken as reference.

The above equations are rigorously valid only inside the crystal. As for X-rays the refraction index of materials is not exactly 1 but slightly smaller; the tiny refraction of the radiation occurring at the vacuum– (or air–) crystal interface should be considered to get accurate energies. This can be done by replacing $d_{hk\ell}$ with an effective spacing constant defined as follows[7]:

$$d_{hk\ell}^{cor} = \left(1 - \frac{\delta}{\sin^2(\vartheta)}\right) \cdot d_{hk\ell} \tag{2.4}$$

where the correction factor $\delta$ is given by:

$$\delta \approx 2.7 \cdot 10^{-6} \cdot \frac{\rho \cdot Z}{M} \cdot \lambda^2, \tag{2.5}$$

with $\rho$ as the crystal density (in g/cm³), $Z$ as the number of electrons in the crystal molecule (or crystal atom) of molar mass $M$ (in g), and $\lambda$ as the wavelength of the radiation (in Å).

The Bragg law is a necessary but not sufficient condition to observe constructive interferences. Actually, the amplitudes and relative phases of the X–rays scattered by the individual atoms in the unit cell of the crystal are given by the structure factor $F_{hkl}$ defined by:

$$F_{hk\ell} = \sum_{j=1}^{N} f_j \cdot e^{2\pi i(h \cdot x_j + k \cdot y_j + \ell \cdot z_j)}, \tag{2.6}$$

where $f_j$ stands for the atomic scattering factor of the $j$th atom whose position in the unit cell is given by the coordinates $x_j$, $y_j$, $z_j$, and $N$ is the number of atoms in the unit cell. The atomic scattering factor $f_j$ corresponds to the ratio between the amplitude scattered by the atom $j$ and the one scattered by an isolated electron, under identical conditions. Clearly, the maximum value that $f_j$ can take is $Z_j$ (atomic number of the atom $j$). When $f_j = Z_j$, all the electrons scatter in phase with each other. As the intensity of a wave is given by the square of its amplitude, the intensity of the photon beam diffracted by the planes ($hk\ell$) is proportional to the square of the modulus of the structure factor $F_{hk\ell}$:

$$I_{hk\ell} \propto \left| F_{hk\ell} \right|^2 = F_{hk\ell}^* \cdot F_{hk\ell}, \tag{2.7}$$

where $F_{hk\ell}^*$ is the complex conjugate of $F_{hk\ell}$. For face-centered cubic crystals like silicon and germanium, the structure factor $F_{hk\ell}$ only when all Miller's indices are odd or all are even. In other words, for face-centered cubic crystals there is no diffracted intensity if the Miller's indices are mixed. For instance, no diffraction is observed for the (110) planes of Ge but a strong diffraction is observed for Ge (220), which is equivalent to Ge (110) in the second order of diffraction. For hexagonal crystals like quartz ($SiO_2$), diffraction by ($00\ell$) planes is only possible when $\ell$ is a multiple of 3.

The Bragg diffraction occurs not only at the Bragg angle but also over a narrow angular range. The width of the rocking curve $I(\vartheta)$ representing the variation of the diffracted intensity around the Bragg angle depends on the crystal type. For perfect crystals such as quartz, silicon, or germanium, the width, usually referred to as the *Darwin width*, is small (see Figure 2.2, dotted curve). According to the dynamical theory,[8–10] the Darwin width $\omega_D$ is proportional to the structure factor and inversely proportional to the energy of the radiation. For mosaic crystals the width of the rocking curve originates predominantly from the mosaicity width $\omega_M$, which corresponds to the FWHM of the crystallites' angular distribution.[11] Depending on the crystal, $\omega_M$ ranges from ~$10^{-4}$ to ~$10^{-2}$ rad.[11,12] For mosaic crystals, the rocking curve results from the convolution of the Darwin profile corresponding to the individual crystallites with the crystallites' angular distribution. If the crystal is thin and both distributions can be represented by Gaussian functions, the convolved width $\omega_{D\otimes M}$ of the rocking curve is given by:

$$\omega_{D\otimes M} = \sqrt{\omega_D^2 + \omega_M^2}. \tag{2.8}$$

For most perfect and mosaic crystals, the rocking curve can be calculated with the computer code X-ray oriented programs (XOP).[13] For illustration, the rocking curves of a perfect Ge (400) and mosaic LiF (420) crystal are presented in Figure 2.2.

The angular resolution of a crystal spectrometer depends not only on the Darwin and mosaicity widths but also on the apparent angular width $\omega_S$ of the radiation source, that is, on the source size, and for spectrometers equipped with a position-sensitive detector on the spatial resolution of the latter, that is, on the detector angular

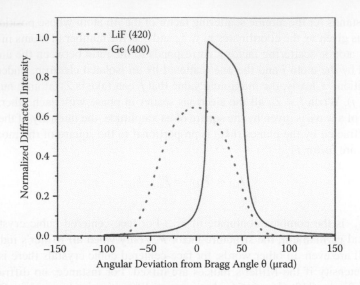

**FIGURE 2.2    (See Color Insert.)** Rocking curves computed with the XOP code[13] for the symmetric Bragg case diffraction of σ-polarized X-rays of 8 keV by a 0.5 mm thick perfect Ge (400) crystal (solid blue curve) and a mosaic (FWHM mosaicity of 0.003 deg.) LiF (420) crystal of the same thickness (dotted red curve). For Ge, as a result of the refraction occurring at the front surface of the crystal, the centroid of the rocking curve is shifted by about 30 μrad from the Bragg angle.

resolution $\omega_d$. The total angular resolution $\omega_\vartheta$ of the spectrometer is obtained by the convolution of the profiles corresponding to each contribution. In the particular case in which all profiles are nearly Gaussian, $\omega_\vartheta$ is therefore given by ($\omega_M = 0$ for perfect crystals):

$$\omega_\vartheta \cong \sqrt{\omega_D^2 + \omega_M^2 + \omega_S^2 + \omega_d^2}. \qquad (2.9)$$

The energy resolution $\omega_E$ of the spectrometer can be derived from the angular resolution $\omega_\vartheta$ using the following equation derived from the Bragg law:

$$\omega_E = ctg(\vartheta) \cdot \omega_\vartheta \cdot E, \qquad (2.10)$$

with $\omega_\vartheta$ in rad.

The width of a measured X-ray line is always bigger than the instrumental resolution $\omega_E$ because the contribution of the natural width characterizing X-ray transitions is not included in Equation (2.10). In this respect, X-ray lines measured with crystal spectrometers are generally fitted with Voigt functions because the latter correspond to the convolution of the Lorentzian natural line shapes of X-ray transitions with the Gaussian instrumental responses of the spectrometers. The width of the Gaussian corresponding to the instrumental broadening of the spectrometer can be directly determined from measurements of γ-rays because the latter have a negligibly small natural width. An alternative solution is to measure a reference X-ray line and to use in the data analysis the Gaussian width as a free fitting parameter while keeping fixed the natural Lorentzian width of the reference line at its known value. If the natural line width of the reference line is not known accurately, the latter can be derived from reliable databases for atomic level widths such as the one of Campbell and Papp.[14]

The reflectivity of a crystal characterizes the fraction of electromagnetic power that is reflected by the diffraction planes. The integral reflectivity is defined by:

$$R_{int} = \int_{\Delta\vartheta} \frac{I(\vartheta)}{I_0} d\vartheta, \tag{2.11}$$

where $I_0$ corresponds to the intensity of the incoming monochromatic radiation and $\Delta\vartheta$ to the angular range around the Bragg angle over which the diffracted intensity is not vanishingly small. The peak reflectivity corresponds to the maximum of the normalized diffracted intensity, that is, to:

$$R_{peak} = \frac{I(\vartheta)_{max}}{I_0}. \tag{2.12}$$

Note that in first approximation the peak and integral reflectivities are related by the following expression:

$$R_{int} \cong \omega_{D\otimes M} \cdot R_{peak}, \tag{2.13}$$

where $\omega_{D\otimes M}$ stands for the width of the rocking curve. For thin crystals, the integral and peak reflectivities can be determined from the kinematical theory[8-10], which is relatively simple but does not take into consideration the interaction between the incoming and outgoing radiation. For thicker crystals, the dynamical theory[8-10] provides more reliable values. According to the kinematical theory, the integral reflectivity varies as $E^{-2}$, whereas an $E^{-1}$ dependence is predicted by the dynamical theory. In both theoretical approaches, the reflectivity depends on the polarization of the incoming X-rays. If the X-rays are linearly polarized and the polarization direction is parallel to the plane of incidence ($\pi$-polarized photons), the reflectivity varies as $\cos^2(2\vartheta)$ (kinematical theory) or $|\cos(2\vartheta)|$ (dynamical theory). As a consequence, for $\pi$-polarized X-rays, the peak and integral reflectivities are nearly zero if the Bragg angle is close to 45 degrees. If the polarization direction is perpendicular to the incidence plane ($\sigma$-polarized photons), the reflectivity is independent from the Bragg angle. For unpolarized X-rays, the average value of the reflectivities corresponding to $\pi$-polarized and $\sigma$-polarized X-rays is used.

The peak and integral reflectivities can also be computed using the XOP code.[13] From the latter, one finds for instance that for the symmetric Bragg diffraction of σ-polarized photons of 8 keV by a 0.5 mm thick LiF (420) crystal having a FWHM mosaicity of 0.003 degree $R_{peak} = 63\%$ and $R_{int} = 55$ µrad (see Figure 2.2, dotted red curve).

## 2.3 BRAGG-TYPE X-RAY CRYSTAL SPECTROMETERS

Bragg crystal spectrometers are classified into plane and curved-crystal spectrometers and can be based on single- or multiple-crystal arrangements. This section gives a description of the principles of the Bragg plane[15] and the curved-crystal spectrometers of Johann,[16] Johansson,[17] and von Hamos[18] geometries, as well as of alternative related designs. It highlights the different designs and performances of a number of existing instruments used worldwide for high-energy resolution XAS/XES spectroscopy in the soft and tender X-ray regimes.

### 2.3.1 BRAGG-TYPE PLANE-CRYSTAL SPECTROMETERS

The simplest design of a Bragg-type X-ray spectrometer is one based on a plane crystal. The crystal and the detector are mounted on a precise goniometer stage, which allows rotating the crystal to a given Bragg angle $\vartheta$ and the detector to twice that angle. X-ray spectra are collected in $\vartheta$–$2\vartheta$ scans. The incident and diffracted beam directions are defined by narrow slits placed between the sample and the crystal and in front of the detector. In the case of a divergent X-ray beam directed upon a flat crystal, only photons fulfilling the Bragg criteria are diffracted. As a consequence only a small zone of the crystal contributes to the diffraction process resulting in a low efficiency of the spectrometer. In addition, X-ray diffraction occurs for slightly different photon energies depending on the direction of photon propagation. Thus, in practice, the relative resolution of the spectrometer is determined by the angular spread of the diffracted beam, which in turn depends primarily on the divergence of the incident radiation and to a smaller extent on the crystal Darwin width.

Thus, in flat crystal spectrometers the radiation incident on the analyzer crystal should be parallel. This can be achieved by inserting between the sample and the crystal a Soller slit collimator (see Figure 2.3) or collimating X-ray optics (half-focusing polycapillary) as shown in Figure 2.4. To preserve the energy resolution, however, the width of the collimator slits should be very small. This in turn strongly reduces the solid angle and, hence, the efficiency of the setup. In the case of collimating X-ray optics devices, the diameter of the transmitted beam cannot exceed about 10 mm so that only a small part of the crystal will contribute to the Bragg diffraction of a particular wavelength. In addition, the polycapillary transmittance is much less than 100% and it varies with the X-ray energy to fall to zero for X-ray energies above about 20 keV.

A number of parallel-beam wavelength-dispersive spectrometers (PBWDSs) based on flat crystals for X-ray diffraction and polycapillary optics have been developed.[19–21] In particular, Szlachetko et al.[22] demonstrated that PBWDSs are well suited for X-ray elemental and chemical analyses using microfocused primary photon

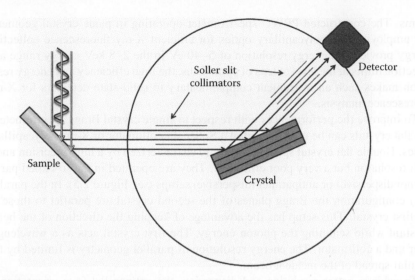

**FIGURE 2.3 (See Color Insert.)** Plane-crystal spectrometer geometry with Soller slit collimators.

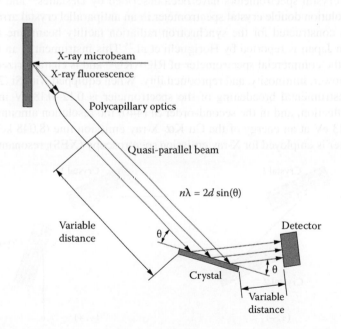

**FIGURE 2.4** Parallel-beam wavelength dispersive arrangement employing polycapillary collimating optics. (Reprinted from Szlachetko, J., Cotte, M., Morse, J. et al., 2010, *J. Synchrotron Rad.*, 17, 400–408.[22] With permission.)

beams. The constructed PBWD spectrometer operating in plane-crystal geometry and employing the polycapillary optics for efficient X-ray fluorescence collection energy provides an energy resolution of 5–40 eV in the 2–8 keV energy range and detection limits at the level of tens of ppm. Thus, the high efficiency and energy resolution makes such an instrument complementary to solid-state detectors for X-ray fluorescence analysis.

To improve the performances with respect to single crystal Bragg spectrometers, two flat crystals can be used instead of a Soller slit collimator or X-ray polycapillary optics. Double flat crystal spectrometers are characterized by a high precision and a high resolution but a very poor efficiency. They are operated in the so-called parallel (nondispersive) or antiparallel (dispersive) setups (see Figure 2.5). In the parallel (+,–) configuration the Bragg planes of the second crystal are parallel to those of the first crystal. This setup has the advantage of keeping the direction of the beam constant while scanning the photon energy. The first crystal acts as a wavelength filter and a collimator. The energy resolution of parallel geometry is limited by the angular spread of the incident beam.

To achieve better resolution and dispersion, the antiparallel (+,+) orientation is preferable. In the dispersive case the reflected wavelength bandwidth after the second crystal is smaller than the one after the first crystal, in contrast to the nondispersive crystal arrangement for which the bandwidth is the same as that after the first crystal.

Double crystal spectrometers have been described by Deslattes[23] and Gohshi.[24] A high-resolution double crystal spectrometer in an antiparallel crystal arrangement which was constructed for the synchrotron radiation facility beam line BL15XU, SPring-8 in Japan is reported by Horiguchi et al.[25] This instrument is an improved version of the commercial spectrometer of RIGAKU[26] and is characterized by high resolving power, luminosity, and reproducibility. When equipped with Si (220) crystals, the instrumental broadening of the spectrometer is 0.12–0.18 eV in the first order of reflection, and in the second-order Si (440) the resolution amounts to only 0.015–0.043 eV at an energy of the Cu $K\alpha_1$ X-ray emission line (8.048 keV).[27] The spectrometer is employed for X-ray emission spectroscopy (XES), resonant inelastic

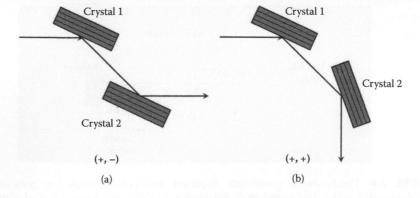

**FIGURE 2.5** Double crystal configurations in Bragg geometry: (a) the nondispersive (+,–), and (b) the dispersive (+,+) arrangement.

X-ray scattering (RIXS), X-ray absorption fine structure spectroscopy (XAFS), and partial fluorescence yield (PFY) spectroscopy.

In order to enhance the energy resolution a four-crystal configuration (+,−,−,+) can be implemented. The first and second pairs of crystals are in the parallel setting, and the two pairs are arranged in the antiparallel configuration. With such a configuration one can achieve a much better energy resolution that is independent of the incident beam divergence, and a more rapid decay of the rocking curve tails.[28] The high-energy resolution is, however, at the severe cost of collection efficiency. The disadvantage of the configuration lies also in the complexity of the design. To simplify the crystals' alignment, channel-cut crystals may be used.

### 2.3.2  Bragg-Type Curved-Crystal Spectrometers

To achieve higher collection efficiency of X-rays from a divergent source, flat crystals are replaced by curved-crystal designs. Since a bent crystal allows using almost the entire crystal surface to diffract X-rays of a given wavelength and bring them to focus, the solid angle of detection increases considerably. In curved-crystal spectrometers, X-ray focusing is achieved either in the dispersive or in the nondispersive plane. In the case of the dispersive (horizontal) focusing, the single or multiple bent crystals are arranged in the so-called Johann[16] or Johansson[17] geometries, and the nondispersive (vertical) focusing corresponds to von Hamos[18] geometry. The crystals are bent cylindrically or spherically. For completion, it should be mentioned that a new type of focusing X-ray spectrometer based on a conically bent crystal was proposed by Hall[29] and elaborated on recently by Morishita, Hayashi, and Nakajima.[30]

#### 2.3.2.1  Johann and Johansson Geometries

The dispersive focusing X-ray curved-crystal spectrometers in which the single or multiple bent crystals are arranged in the Johann[16] or Johansson[17] geometries are Rowland circle-based instruments. The principles of X-ray focusing of diffracted X-ray radiation in the Johann and Johansson geometries are illustrated in Figure 2.6. The conditions which need to be satisfied to achieve diffraction and focusing with bent crystals were first demonstrated by DuMond and Kirkpatrick in 1930.[31] More details and a review on early developments of curved-crystal spectrometers can be found in the articles of DuMond and Chap[32] and Knowles and Chap.[33]

The focusing principle for reflection geometry is that the crystal should be bent to a radius 2R, where R is the radius of the Rowland circle. The Rowland circle can be thus defined as the focal circle whose diameter equals the bending radius of atomic planes of the crystal analyzer. For point-to-point focusing, as shown in Figure 2.6a, the atomic planes of a single crystal should be bent so that they coincide with concentric circles centered at point C, and the crystal boundary surface is ground to a radius R.

Johansson constructed the first curved-crystal spectrometer following this scheme in 1933[17] and since that time, this configuration has been called *Johansson geometry*. For a thin crystal, this geometry results in focusing over the entire range of angles, and almost the whole crystal surface contributes to the diffraction. The different

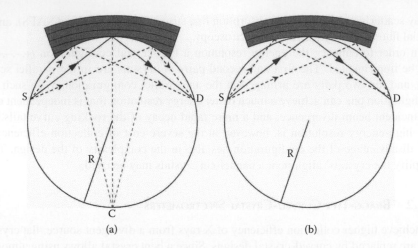

(a)    (b)

**FIGURE 2.6**    (a) Johansson and (b) Johann curved-crystal spectrometer geometries.

X-ray energies from a point source S are brought to a focus on points lying symmetrically at about the center of the crystal surface on the Rowland circle at D. Thus, a spectrometer in Johansson geometry is an exact focusing instrument.

In the case when the crystal planes are bent as in Johansson geometry but the crystal inner surface is not ground, the geometry is semifocusing and is known as *Johann geometry*[16] (see Figure 2.6b). The approximate focusing leading to geometrical aberrations is due to the fact that the extremities of the bent crystal deviate from the focal circle. When small crystal lengths are used, the Johann design is in most cases sufficient. The effect of the different geometrical aberrations on the energy resolution due to the finite size of the crystal and the source, and the bending radius have been investigated extensively and reported by Zschornack, Müller, and Musiol,[34] Beyer and Liesen,[35] Chantler and Deslattes,[36] Chantler,[37] and Hague and Laporte.[38]

The Johann and Johansson curved-crystal spectrometers based on Rowland circle geometry are by definition point-to-point focusing instruments. For a fixed bending radius of typically 0.5–1 m, the distances from the crystal to the source and from the crystal to the detector should be equal. Thus, for a stationary source, collecting an X-ray spectrum requires scanning both the crystal and the detector. When the Bragg angle is changed by rotating the crystal, the source to crystal distance changes, and thus the detector position has to be adjusted accordingly. The choice of the crystal size, diffraction order, and the bending radius is a compromise between higher resolving power and loss of luminosity. For Johann-type spectrometers, smaller crystals and backscattering geometries yield higher-energy resolution.

With the advent of synchrotron radiation facilities providing intense and focused photon beams, as well as the development of 2D position-sensitive detectors and progress in bent crystal manufacturing (bonding of large perfect crystal wafers to glass substrates, or diced and segmented crystals), numerous curved-crystal X-ray spectrometers of Johann and Johansson type have been constructed. Although relying on the Rowland circle focusing principle, many of the spectrometer designs are innovative with respect to the standard Johann and Johansson geometries. In the following some of these instruments are highlighted.

Several multicrystal Johann-type point-by-point scanning spectrometers equipped with spherically bent crystals arranged on intersecting Rowland circles to increase the solid angle of collection, and operating in backscattering geometry and high orders of diffraction for high resolving power of ~10000 and reaching meV energy resolution have been built and are in operation at synchrotron radiation facilities.[39–53] The concept of overlapping Rowland circles with spherically bent crystals is illustrated in Figure 2.7. These spectrometers are employed for XES, RIXS, resonant Raman X-ray scattering (RRS), and PFY X-ray absorption spectroscopy studies typically in the ~5–18 keV range.

An alternative concept for a bent-crystal spectrometer is based on Johann geometry with a variable crystal bending radius, a source-to-crystal distance is fixed and the detector inside the Rowland circle was implemented by Journel.[54] This UHV spectrometer is dedicated for the 1–8 keV energy range with a resolving power of ~7500 and is adapted for gas-phase samples. A Johann-type curved-crystal spectrometer for the 2–5 keV energy range allowing detection of emitted X-rays with variable polarization from a gas-phase target by rotating the spectrometer by 180 degrees was reported by Hudson et al.[55] The sample cell location inside the Rowland circle creates an extended virtual source, and a position-sensitive 2D detector on the Rowland

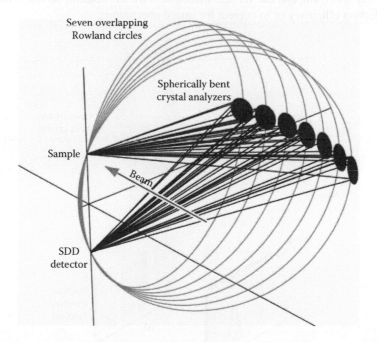

**FIGURE 2.7** **(See Color Insert.)** Overlapping Rowland circles concept for a multicrystal Johann-type spectrometer. (Reprinted from Sokaras, D. et al., 2013, *Rev. Sci. Instrum.*, 84, 053102.[39] Copyright 2013, AIP Publishing LLC. With permission.)

circle allows data collection over a ~20 eV energy range with a resolution of 0.5 eV at the Cl K-edge. A versatile Johansson-type spectrometer in the off-Rowland circle geometry for RIXS and XES spectroscopy in the ~2–6 keV energy range and a sub-natural line width energy resolution was designed by Kavčič et al.[56]

Recently, the so-called short (1–3 cm) working distance (SWD) or miniature (miniXES) compact X-ray spectrometers have been developed. A Johansson-type spectrometer for an energy range of about 3 to 10 keV and collection solid angle ~50 mSr, roughly equivalent to six spherically bent crystals was constructed by Pacold.[57] An alternative design based on millimeter-sized flat analyzer crystals arranged on a spherical surface in short distance from a microfocused source and a 2D position-sensitive detector is described by Dickinson.[58]

### 2.3.2.2 Von Hamos Geometry

The principle of a Bragg spectrometer in von Hamos geometry[18] is shown schematically in Figure 2.8. The crystal is bent cylindrically in the direction perpendicular to the dispersive direction and thus provides focusing in the nondispersive plane. In this arrangement, the X-ray fluorescence source and the position-sensitive detector are located on the axis of dispersion (crystal axis of curvature), and the diffracted X-rays form a two-dimensional image on the detector. The location of the image on the dispersion axis of the detector corresponds geometrically to the wavelength/energy axis of an X-ray spectrum, and the vertical direction of the 2D detector serves to increase the collection efficiency or to correct for image aberrations.

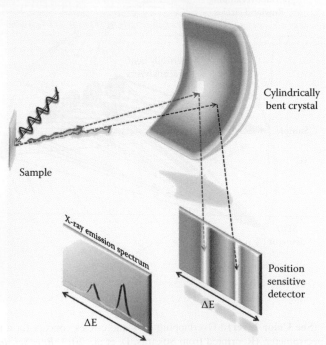

**FIGURE 2.8   (See Color Insert.)** Schematic view of von Hamos geometry.

The advantage of vertical focusing is that it allows the recording of an X-ray emission spectrum over a certain energy bandwidth. The latter is determined by the crystal and detector lengths along the dispersion access. The distance along the dispersion axis between the crystal center and the fixed source is $R \cdot ctg\vartheta$, where $R$ is the crystal radius of curvature and $\vartheta$ the Bragg angle. The detector distance is twice that of the crystal, that is, $2R \cdot ctg\vartheta$. To change the angular domain, the crystal and the detector are translated along their axes. The Bragg relation and the angular range of the spectrometer determine the energy range that can be measured with a given crystal. To cover a wider energy range different crystals need to be employed.

In von Hamos geometry, the energy resolution and the efficiency of the X-ray spectrometer depend on the crystal quality and bending radius, the source size, and the detector spatial resolution in the dispersive plane. The larger the radius of curvature, the higher is the energy resolution but the collection efficiency decreases as the inverse power of the radius. To increase the solid angle, larger area crystals may be used, however, crystal bending in the nondispersive plane also influences the diffraction properties. The detector spatial resolution should be matched to the X-ray fluorescence source size, which is determined either by the beam spot dimensions or by a rectangular slit.

In the following, the main characteristics and performances of several existing von Hamos X-ray spectrometers are presented. A versatile high-energy resolution von Hamos instrument for XES and RIXS spectroscopy was constructed at the University of Fribourg, Switzerland, by Hoszowska ct al.[59,60] Thanks to slit geometry, X-ray emission spectra induced by photon and charged particle beams, both in-house with conventional sources and at synchrotron or particle beam facilities can be measured (see Figure 2.9). The spectrometer may be operated in the slitless mode using focused beams, thus allowing grazing emission X-ray fluorescence (GEXRF) measurements.[61] XAS spectra collection with laboratory X-ray sources is also possible. To cover the wide 0.8–16.8 keV energy domain and to optimize the energy resolution, 10 interchangeable crystals bent cylindrically to a radius of 25.4 cm (see Figure 2.10) are employed. For X-ray detection the spectrometer is equipped with either a back- or front-illuminated CCD (charge-coupled device) 2D detector.[62] Depending on the Bragg angle the energy bandwidth spans from 30 eV to 300 eV. The resolving power is typically ~3000 and can be improved by using narrower slits, microfocused photon beams, or higher orders of diffraction.

In von Hamos geometry, high collection efficiency, however, at the expense of energy resolution may be achieved with miniXES spectrometer designs[63,64] or by using mosaic crystals.[65,66] To increase the solid angle and preserve the high-energy resolution, multiple-crystal arrangements in von Hamos geometry were proposed (see Figure 2.11). Hayashi et al.[67] used the concept of five crystals, bent to a radius of 55 cm, set to focus on different detector positions in the nondispersive plane for XES and RIXS spectra detection. A new plastic deformation technique for strongly and accurately shaped Si and Ge wafers was also developed.[68] To combine a large solid angle with sub eV energy resolution for XES and near-edge XRS in the 5–10 keV energy region, a von Hamos spectrometer based on an array of 16 crystals bent

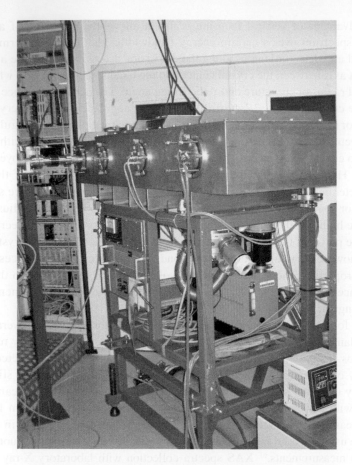

**FIGURE 2.9    (See Color Insert.)** The von Hamos X-ray spectrometer from the University of Fribourg installed at the PHOENIX beam line, Swiss Light Source (SLS), Villigen, Switzerland.

to a radius of 50 cm was constructed by Alonso-Mori et al.[69] This instrument was employed recently at the Linac Coherent Light Source (LCLS) for simultaneous X-ray diffraction and X-ray emission studies of a Photosystem II at room temperature.[70]

A von Hamos spectrometer based on segmented-type crystal diffraction was designed by Szlachetko et al.[71] and Szlachetko et al.,[72] and installed at the SuperXAS beam line at SLS, Switzerland (see Figure 2.12). The use of a segmented crystal (see Figure 2.13) allows preserving the spectrometer efficiency and energy resolution of 0.25 to 1 eV in the 8000–9600 eV energy range. The 25 cm radius for the crystal provides the required resolving power and collection efficiency, and ensures that there is enough space in the sample environment for employing high-temperature and high-pressure cells. The design of the spectrometer allows for different multiple-crystal configurations when using a 2D position-sensitive detector. The instrument can be applied for single-shot XES and time-resolved RIXS studies for many applications and *in situ* spectroscopic studies of dynamic systems.

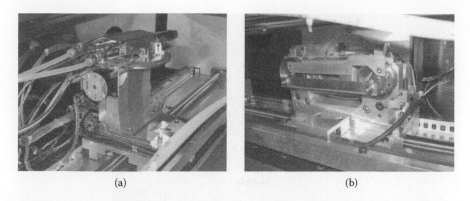

(a)                                                              (b)

**FIGURE 2.10** **(See Color Insert.)** Crystal lamina permanently glued to an aluminum block machined to a precise concave cylindrical surface of a nominal radius of 25.4 cm (a), and the CCD 2D detector (b).

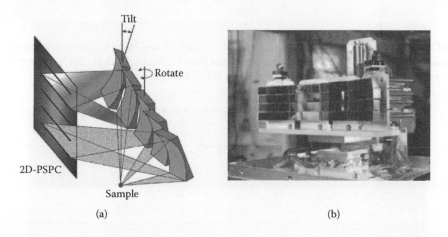

(a)                                                              (b)

**FIGURE 2.11** **(See Color Insert.)** Multiple crystals in von Hamos geometry. (a) Schematic of a five-crystal spectrometer and a 2D position-sensitive detector. (Reprinted from Hayashi, H. et al., 2004, *J. Electron Spectrosc. Relat. Phenom.*, 136, 191.[67] Copyright 2004. With permission from Elsevier.) (b) A picture of the 4 × 4 array of analyzer crystals in von Hamos geometry. (Reprinted from Alonso-Mori, R. et al., 2012, *Rev. Sci. Instrum.* 83, 073114.[69] Copyright 2012, AIP Publishing LLC. With permission.)

## 2.4 LAUE-TYPE X-RAY CRYSTAL SPECTROMETERS

In contrast to the Bragg case, in Laue geometry, the measured radiation has to cross the crystal thickness before reaching the detector. The resulting intensity attenuation which depends on the energy of the photons can be considerable for X-rays below ~15 keV. A further reduction of the intensity in the Laue case arises from the peak reflectivity. For zero absorption and thick crystals, the peak reflectivity is indeed limited to 50% in the Laue case, whereas it can reach 100% in the Bragg case.[73] Most

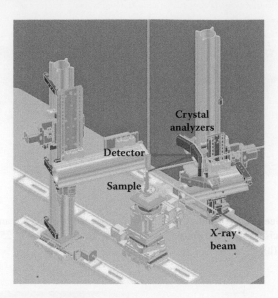

**FIGURE 2.12    (See Color Insert.)** Schematic drawing of von Hamos spectrometer developed at the SLS SuperXAS beam line. (Reprinted from Szlachetko, J. et al., 2013, *J. Electron Spectrosc. Rel. Phenom.,* 188, 161.[72] Copyright 2004. With permission from Elsevier.)

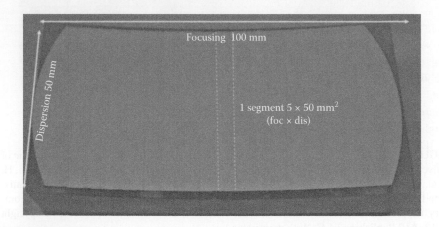

**FIGURE 2.13    (See Color Insert.)** A segmented Si(111) crystal bent to radius of 25 cm. (Reprinted from Szlachetko, J. et al., 2012, *Rev. Sci. Instrum.* 83, 103105.[71] With permission. Copyright 2012, AIP Publishing LLC.)

Laue-type X-ray crystal spectrometers are therefore used for measurements above 15–20 keV and only thin crystals made of light elements such as, for example, LiF, Beryl, $SiO_2$, and Si are employed. Because in the Laue case the detector is located behind the crystal, transmission-type spectrometers need in general more space than Bragg-type instruments, which makes their installation at beam lines of large facilities more difficult. As in the Bragg case, transmission-type instruments can be classified into plane (or flat) crystal spectrometers and curved-crystal spectrometers.

## 2.4.1 Laue-Type Plane-Crystal Spectrometers

Setups similar to those discussed in the section concerning the Bragg-type plane-crystal spectrometers are used, except that in transmission geometry, the radiation source and the detector are positioned on opposite sides of the mono-crystal or double-crystal analyzer. An example of a flat crystal spectrometer constructed to measure γ-rays with energies ranging from 60 keV to 5 MeV is given in Knowles.[74]

As in the Bragg case, for setups based on a single crystal, the divergent X-ray radiation is first transformed into a narrow quasi-parallel beam by means of a Soller slit collimator or a half-focusing polycapillary X-ray optics. The diffracted X-rays are measured by an energy dispersive detector, which is aligned on the double Bragg angle with respect to the direction of the beam impinging on the crystal. The spectral analysis of the radiation is performed by making step-by-step scans of the angular ranges of interest, using $\vartheta$–$2\vartheta$ rotary stages. To reduce the background events, a second Soller slit collimator is generally installed on the $2\vartheta$ stage between the crystal and the detector. Because only a small part of the crystal contributes to the diffraction, such spectrometers have a low luminosity. Irradiating a larger area of the crystal surface with a divergent beam from a point-like source and using a 2D position-sensitive detector can enhance the luminosity. This permits the simultaneous measurement of a certain wavelength region, X-rays with different wavelengths being measured at different positions over the whole detector width. If the diffraction planes are vertical, X-ray transitions with different energies correspond to separate vertical lines on the 2D detector.

As the Bragg angle varies smoothly with the height of the impact point of the photon on the crystal, the vertical lines can be slightly curved. The solid angle and consequently the luminosity of the setup grows when the distance $\ell$ between the source and the crystal decreases. On the other hand, the optical aberrations diminish with increasing values of $\ell$ so that a compromise should be made to get simultaneously a high enough luminosity and acceptable geometrical aberrations.

Double flat crystal spectrometers have been also developed for the Laue diffraction case.[23,75–77] They are characterized by an ultra-high resolving power $\Delta E/E$ of $10^{-4}$–$10^{-6}$ but also, due to the double diffraction, by a very low overall efficiency of the order of $10^{-11}$, which is several orders of magnitude smaller than the one corresponding to setups based on a single plane crystal. For this reason they are mainly used for high-resolution measurements of intense high-energy γ-rays (see, for example, Stritt et al.[78]) and to a smaller extent for the metrology of X-ray transitions (see Deslattes et al.[79] and Kessler et al.[80]). Double flat Laue-type crystal spectrometers can be operated in the nondispersive (n,–n) mode or in the dispersive (n,m) mode with m = n or any other diffraction order $m \neq n$ for which a reflection can be observed (see Figure 2.14). In the nondispersive or parallel mode, the beam, which exits from the spectrometer, is parallel with the beam that enters the spectrometer. There is no deflection of the beam and thus no angular dispersion. For the dispersive or antiparallel mode, the angular dispersion is given by the sum of the dispersion corresponding to each crystal. The instrumental response of the spectrometer can be determined directly from measurements performed in the antiparallel mode,

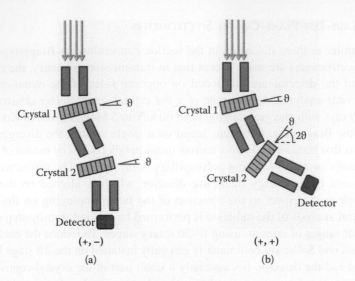

**FIGURE 2.14    (See Color Insert.)** (a) Parallel nondispersive (+,-) and (b) antiparallel dispersive (+,+) setups of a Laue double flat crystal spectrometer.

broadening effects such as the divergence of the incoming beam being automatically canceled in this geometry.

## 2.4.2    LAUE-TYPE CURVED-CRYSTAL SPECTROMETERS

Most curved-crystal spectrometers based on the Laue diffraction are designed in the line source and line focus geometries. Some alternative geometries such as the so-called Focusing Compensated Asymmetric Laue (FOCAL) setup are also employed for dedicated applications.

### 2.4.2.1    Line Source Geometry

The main geometrical features of a line source transmission spectrometer are depicted in Figure 2.15. The thin crystal lamina is curved cylindrically with a radius of curvature equal to the diameter of the focal circle passing through the crystal center C and curvature center $C_1$. If the line source is located on the focal circle, all photons impinging on the crystal make the same angle $\vartheta$ with the diffraction planes, provided that the dimensions of the crystal lamina remain small as compared to the radius of curvature. If the wavelength of the photons and the angle $\vartheta$ satisfy the Bragg law, the photons are diffracted by the crystal and detected by the extended detector D. Each different position of the source on the focal circle corresponds to a different Bragg angle and thus to a different wavelength. Thus, in this geometry the bent crystal acts as a monochromator. Because of the close angular proximity of the direct radiation at small Bragg angles, that is, at high energies, it is necessary to shield the detector. This is achieved by inserting a Soller slit collimator between the crystal and the detector.

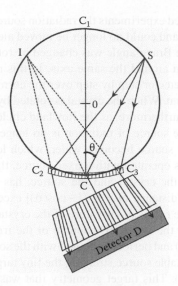

**FIGURE 2.15**  A Laue spectrometer in the source line (DuMond) geometry.

DuMond developed the first line source transmission-type bent crystal spectrometer in 1947.[81] In the original DuMond spectrometer a $80 \times 70 \times 1$ mm$^3$ SiO$_2$ (310) crystal plate was bent to a radius of 2 m. The useful aperture in the crystal holder had an area of 10 cm$^2$ and subtended a solid angle of $2.5 \cdot 10^{-4}$ sr at the focus. The Soller slit collimator placed in front of the detector allowed DuMond to extend the measurements to small Bragg angles, that is, to measure high-energy $\gamma$-rays up to 1.75 MeV. The Soller slit collimator and the shielding surrounding the detector were much heavier than the radioactive source. For this reason, in the original setup of DuMond, the heavy collimator-detector system remained stationary and the radiation source and diffracting crystal were rotated about a vertical axis passing through point C to vary the Bragg angle.

Since this pioneering work, many spectrometers based on DuMond geometry have been constructed.[82–92] Most of them were installed at beam lines of large facilities for high-resolution (n,$\gamma$) spectroscopy[93] or for the investigation of ($\alpha$,xn) nuclear reactions[94] and heavy-ion induced, multiple inner-shell atomic ionization[95–97] or for the measurement of X-rays from muonic[98] and pionic atoms.[99] More recently, a multitude of XES (X-ray emission spectroscopy) and RIXS (resonant inelastic X-ray scattering) experiments have been carried out at synchrotron radiation (SR) facilities and X-ray free electron laser (XFEL) sources using crystal spectrometers. Due to the limited space available at the SR and XFEL beam lines, most measurements have been performed with Bragg-type curved crystal spectrometers, which are more compact. However, a few X-ray absorption spectroscopy experiments have also been carried out using Laue-type analyzers equipped with bent crystals characterized by short focal distances.[100–104]

In all the aforementioned experiments the radiation source was a target that had to be kept fixed in the beam and could no longer be moved along the Rowland circle to vary the Bragg angle. The Bragg angle was changed by rotating the crystal and the collimator-detector system around the same axis. In this mode of operation, scanning the crystal and the detector step by step over the corresponding angular region collects an energy spectrum. When the crystal is rotated by an angle $\vartheta$, the detector has to be rotated by $2\vartheta$. Furthermore, as the Rowland circle rotates with the crystal, after a crystal rotation the source of radiation is no longer on the focal circle and the apparent width of the source becomes wider, which leads to resolution losses. In DuMond spectrometers operated with a fixed source, the focusing distance, that is, the distance between the crystal and the source, has therefore to be adjusted whenever the defocusing distance $\delta f = R \cdot [1 - \cos(\vartheta)]$ exceeds a value of the order of $\sim R/1000$, where $R$ is the radius of curvature of the crystal.

In DuMond geometry, the radioactive source or the irradiated target should be very thin ($\sim 0.01$–$0.10$ mm) and perfectly aligned with the source-crystal direction. In order to achieve an acceptable source strength, the tiny target width is compensated by a large depth ($\sim 5$ mm). This target geometry that was introduced by DuMond represents still the standard geometry for $\gamma$-ray measurements. For X-rays, however, whose energies are in general lower than those of $\gamma$-rays, this geometry is less appropriate because the X-rays produced in the depth of the target are almost fully absorbed by the target itself. Therefore, for X-ray spectroscopy measurements, the spectrometer is usually operated in the so-called modified DuMond slit geometry.[91] In this geometry, represented schematically in Figure 2.16, a narrow rectangular slit is placed on the focal circle at a fixed position and serves as the effective source of radiation. The target placed behind the slit can be tilted around a vertical axis to make a certain angle with respect to the target-crystal direction. For a given X-ray line, the angle is chosen so that the highest counting rate is obtained in the spectrometer detector. The variation of the counting rate as a function of the target alignment depends on the self-absorption of the X-rays in the target and the size of the target surface viewed by the crystal through the slit. Both quantities decrease with growing angles but with opposite effects on the counting rate. Two further advantages are provided by this geometry: first, the slit width can be adjusted to obtain the desired instrumental resolution, and second, the line shapes of the measured X-rays are no more affected by thermal deformations of the target.

Recent spectrometers are operated in a fully automatic way by computers, which control also the data acquisition. Regarding the crystals, the know-how acquired at synchrotron radiation facilities in the preparation of ultra-pure crystals and the development of new bending devices have also been of great benefit. However, despite these improvements, the basic mechanical design and the working principles of modern line source spectrometers have remained essentially the same as those of the DuMond prototype.

### 2.4.2.2   Line Focus Geometry

As shown in Figure 2.17, the geometry of the line focus transmission spectrometer is similar to the one of the DuMond setup except that the source and detector are interchanged. This design permits the use of an extended source and focuses the radiation

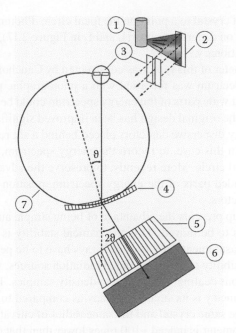

**FIGURE 2.16** **(See Color Insert.)** Schematic drawing of the modified DuMond slit geometry: (1) X-ray tube, (2) target, (3) slit, (4) cylindrically bent crystal, (5) Soller slit collimator, (6) detector, and (7) focal circle.

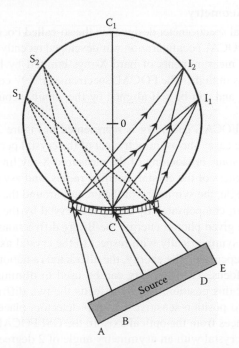

**FIGURE 2.17** **(See Color Insert.)** Laue spectrometer in line focus (Cauchois) geometry.

diffracted by the bent crystal to a point on the focal circle. Photons of different energies are thus focused on different points ($I_1$ and $I_2$ in Figure 2.17), the source and the crystal remaining stationary.

The first spectrometer of this type was constructed by Cauchois.[105] In the original design the energy spectrum was recorded with a photographic plate positioned on the focal circle so that wide parts of the energy spectrum could be measured simultaneously. Since then, the original design has been improved significantly.[106–108] In particular, various energy dispersive detectors placed behind a slit replaced the original photographic plate. In this case, to record the energy spectrum, the slit detector is moved along the focal circle. More recently, to preserve the advantage of collecting simultaneously extended parts of the energy spectrum, position-sensitive detectors replaced the slit detectors.

The Cauchois setup presents the advantage of being simple and very stable since only the detector has to be moved. This mechanical stability is very useful whenever very precise measurements of γ-rays or X-rays have to be performed. A further advantage is the possibility of using extended radiation sources, which is very helpful in some applications dealing with very low-density samples. The main drawback of the line focus geometry is its small efficiency as compared to that of line source spectrometers. For the same crystal and the same radius of curvature, the luminosity of a Cauchois spectrometer is indeed ~100 times lower than that of a DuMond spectrometer. This drawback, however, can be partly compensated by using an extended position-sensitive detector.

### 2.4.2.3   FOCAL Geometry

A transmission crystal spectrometer designed in the so-called Focusing Compensated Asymmetric Laue (FOCAL) configuration was developed recently at GSI, Darmstadt, for accurate energy measurements of hard X-rays emitted by heavy one-electron and few-electron ions in flight. The FOCAL spectrometer[109–111] covers a wide energy range (30–120 keV) and can be self-aligned by the simultaneous double measurement of X-ray lines.

The principle of FOCAL geometry is represented in Figure 2.18. It is based on the asymmetric Laue case. The asymmetric cut of the crystal permits increasing the reflectivity but with some broadening of the measured X-ray lines. The broadening depends on the thickness of the crystal, bending radius, and asymmetry angle. Due to the asymmetry angle, the symmetry of the images around the crystal axis is lost. This can be, however, compensated by tilting the crystal by the same angle. In the FOCAL setup, for a given photon energy, the Bragg diffractions occur at two narrow regions located symmetrically with respect to the crystal axis. While the position of the regions depends on the energy, the diffracted radiation passes a common polychromatic line focus. This property can be used to diminish the background by inserting a slit at this position. After the focus the two diffracted X-ray beams are measured by two position-sensitive Ge strip detectors placed on the Rowland circle at equal distances from the optical axis. In the GSI FOCAL setup, a 120 × 40 × 1.5 mm³ Si (220) crystal with an asymmetry angle of 2 degrees and a cylindrical

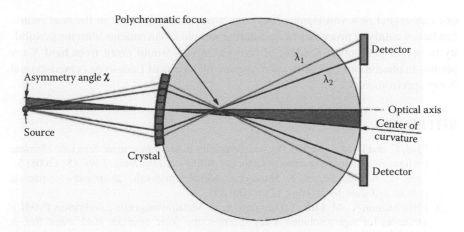

**FIGURE 2.18** **(See Color Insert.)** Schematic drawing showing the principle of FOCAL geometry. (From Beyer, H.F. et al., 2009, *Spectrochim. Acta B* 64, 736.[111] Copyright 2009. With permission from Elsevier.)

radius of curvature of 200 cm is used. The spectrometer efficiency was estimated to be $1.2 \times 10^{-7}$ for a detector width of 100 mm and the FWHM energy resolution was found to be less than 100 eV at about 50 keV.

## 2.5 SUMMARY AND CONCLUDING REMARKS

It is now one century since the pioneering work of Bragg in 1913 that X-ray crystal spectrometers were constructed for high-resolution X-ray spectroscopy. The availability of high brilliance synchrotron radiation facilities and femtosecond XFEL sources has given a boost to the domain, and one can expect to witness further progress in both X-ray crystal spectrometer development and related applications. The objective of this chapter was to give an overview of the basic principles of Bragg- and Laue-type geometries for plane- and curved-crystal X-ray spectrometers, as well as to highlight the different designs and performances worldwide of operating instruments dedicated to high-energy resolution X-ray photon-in photon-out core level spectroscopy.

Understanding catalytic performances requires mapping of the electronic structure of the catalyst under reaction conditions. In this respect, the nondispersive focusing and dispersive diffraction allowing collection of high-resolution X-ray emission spectra on a shot-by-shot basis with an energy range from few tens to few hundreds of eV is of particular relevance for time-resolved studies of catalytic reactions under different *in situ* conditions. The versatility of von Hamos geometry permits optimizing the spectrometer design and performance in terms of resolving power, collection efficiency, and energy range for very different applications using synchrotron and XFEL sources. In this perspective, although to date in comparison to the Rowland circle-based curved-crystal spectrometers, the von Hamos-type instruments were less frequently used for XES/XAS spectroscopy in the tender X-ray energy range, one can expect new von Hamos X-ray spectrometers to be built in the near future. Studies of catalytic processes in demanding sample environments with the possibility to access absorption K-edges of heavy elements would profit from hard X-ray photon-in photon-out core level spectroscopy by means of Laue-type curved-crystal X-ray spectrometers.

# REFERENCES

1. Enss, C. and D. McCammon. Physical principles of low temperature detectors: Ultimate performance limits and current detector capabilities. *J. Low Temp. Phys.* 151 (2008) 5.
2. Enss, C., A. Fleischmann, K. Horst et al. Metallic magnetic calorimeters for particle detection. *J. Low Temp. Phys.* 121 (2000) 137.
3. Fleischmann, A., M. Link, T. Daniyarov et al. Metallic magnetic calorimeters (MMC): Detectors for high-resolution X-ray spectroscopy. *Nucl. Instrum. Meth. Phys. Res.* A 520 (2004) 27.
4. Lind, D.A., W.J. West, and J.W.M. DuMond. X-ray and gamma-ray reflection properties from 500 X units to nine X units of unstressed and of bent quartz plates for use in the two-meter curved-crystal focusing gamma-ray spectrometer. *Phys. Rev.* 77 (1950) 475.
5. Mohr, P.J., B.N. Taylor, and D.B. Newell. CODATA recommended values of the fundamental physical constants: 2006. *J. Phys. Chem. Ref. Data* 37 (2008) 1187.
6. Deslattes, R.D., E.G. Kessler, P. Indelicato, L. de Billy, E. Lindroth, and J. Anton. X-ray transition energies: New approach to a comprehensive evaluation. *Rev. Mod. Phys.* 75 (2003) 35.
7. Zschornack, G. In *Handbook of X-Ray Data,* Berlin, Heidelberg: Springer-Verlag, 2007, p. 129.
8. James, R.W. In *The Crystalline State, Vol. 2, The Optical Principles of the Diffraction of X-Rays*, Chapter II: The Intensity of Reflection of X-Rays by Crystals. London: G. Bell and Sons, 1967.
9. Azaroff, L.V. In *Elements of X-Ray Crystallography*, Chapter 9: X-ray Diffraction by Ideal Crystals. New York: McGraw-Hill, 1968.
10. Azaroff, L.V., R. Kaplow, N. Kato, R.J. Weiss, A.J.C. Wilson, and R.A. Young. In *X-Ray Diffraction,* Chapter 2: Kinematical Theory and Chapter 4: Dynamical Theory for Perfect Crystals. New York: McGraw-Hill, 1974.
11. Darwin, C.G. The reflexion of X-rays from imperfect crystals. *Phil. Mag.* 43 (1922) 800.
12. Renninger, M. Studien über die Röntgenreflexion an Steinsalz und den Realbau von Steinsalz *Z. Kristallogr. Z. Phys.* 89 (1934) 344.
13. Sanchez del Rio, M. *X-ray Oriented Programs (XOP)*, http://www.esrf.eu/computing/scientific/xop2.0/.
14. Campbell, J.L. and T. Papp. Widths of the atomic $K-N_7$ levels. *At. Data Nucl. Data Tables* 77 (2011) 1.
15. Bragg, W.H. and W.L. Bragg. The deflection of X-rays by crystals. *Proc. Roy. Soc.* 88A (1913) 428.
16. Johann, H.H. Die Erzeugung lichtstarker Rontgenspektren mit Hilfe von Konkavkristallen. *Z. Physik* 69 (1931) 185.
17. Johansson, T. Über ein neuartiges, genau fokussierendes Röntgenspektrometer. *Z. Physik* 82 (1933) 507.
18. Von Hamos, L. Roentgenspektroskopie und Abbildung mittels gekruemmter Kristallreflektoren. *Naturwiss.* 20 (1932) 705.
19. Soeima, H. and T. Narusawa. A compact X-ray spectrometer with multi-capillary X-ray lens and flat crystals. *Adv. X-ray Anal.* 44 (2001) 320.

20. Schields, P.J., D.M. Gibson, W.M. Gibson, N. Gao, H. Huang, and I. Yu. Ponomarev. Overview of polycapillary X-ray optics. *Powder Diffr.* 17 (2002) 70.
21. Van Hoek, C. and M. Koolwijk. Conventional wavelength dispersive spectroscopy versus parallel beam spectroscopy—A basic overview. *Microchim. Acta* 161 (2008) 287.
22. Szlachetko, J., M. Cotte, J. Morse et al. Wavelength-dispersive spectrometer for X-ray microfluorescence analysis at the X-ray microscopy beamline ID21 (ESRF). *J. Synchrotron Rad.* 17 (2010) 400.
23. Deslattes, R.D. Two-crystal, vacuum monochromator. *Rev. Sci. Instrum.* 38 (1967) 616.
24. Gohshi. Y. , H. Kamada, K. Kohra, T. Utaka and T. Arai. Wide range two-crystal vacuum X-ray spectrometer for chemical state analysis, *Appl. Spectr.* 36 (1982) 171.
25. Horiguchi, D., K. Yokoi, H., Mizota et al. Anti-parallel crystal spectrometer at BL15XU in SPring-8, first results. *Rad. Phys. Chem.* 75 (2006) 1830.
26. Gohshi, Y., H. Kamada, K. Kohra, T. Utaka, and T. Arai. Optics InfoBase applied spectroscopy—Wide-range two-crystal vacuum X-ray spectrometer for chemical state analysis. *Appl. Spectrosc.* 36 (1982) 171.
27. Tochio, T., Y. Ito, and K. Omote. Broadening of the X-ray emission line due to the instrumental function of the double-crystal spectrometer. *Phys. Rev. A* 65 (2002) 042502.
28. Sutter, J.P., G. Duller, S. Hayama, U. Wagner, and S. Diaz-Moreno. Performance of multi-crystal Bragg X-ray spectrometers under the influence of angular misalignments. *Nucl. Instrum. Methods Phys. Res. A* 589 (2008) 118.
29. Hall, T.A. A focusing X-ray crystal spectrograph. *J. Phys. E* 17 (1984) 110.
30. Morishita, K., K. Hayashi, and K. Nakajima. One-shot spectrometer for several elements using an integrated conical crystal analyzer. *Rev. Sci. Instrum.* 83 (2012) 013112.
31. DuMond, J.W.M. and H.A. Kirkpatrick. The multiple crystal X-ray spectrograph. *Rev. Sci. Instrum.* 1 (1930) 88.
32. DuMond, J.W.M. In *Beta and Gamma-ray Spectroscopy*, 1st ed., Chapter IV: Crystal Diffraction Spectroscopy of Nuclear γ-Rays, K. Siegbahn, ed. Amsterdam: North-Holland, 1955.
33. Knowles, J.W. Chap. IV in *Beta and Gamma-Ray Spectroscopy*, 2nd ed., Chapter IV: Crystal Diffraction Spectroscopy of Nuclear γ-Rays, K. Siegbahn ed. Amsterdam: North-Holland, 1965.
34. Zschornack, G., G. Müller, and G. Musiol. Geometrical aberrations in curved Bragg crystal spectrometers. *Nucl. Instrum. Meth. A* 200 (1982) 481.
35. H. Beyer and D. Liesen. Ray tracing of curved-crystal X-ray optics for spectroscopy on fast ion beams. *Nucl. Instrum. Meth. A* 272 (1988) 895.
36. Chantler, C.T. and R.D. Deslattes. Systematic corrections in Bragg X-ray diffraction of flat and curved crystals. *Rev. Sci. Instrum.* 66 (1995) 5123.
37. Chantler, C.T. X-ray diffraction of bent crystals in Bragg geometry. I. Perfect-crystal modelling. *J. Appl. Cryst.* 25 (1992) 674.
38. Hague, C.F. and D. Laporte. Spectrometer for soft X-ray emission studies of liquid metals and metallic vapors. *Rev. Sci. Instrum.* 51 (1980) 621.
39. Sokaras, D., T.-C. Weng, D. Nordlund et al. A seven-crystal Johann-type hard X-ray spectrometer at the Stanford Synchrotron Radiation Lightsource. *Rev. Sci. Instrum.* 84 (2013) 053102.
40. Bergmann, U. and P. Glatzel. X-ray emission spectroscopy. *Photosynth. Res.* 102 (2009) 255.
41. Kao, C.C., K. Hämäläinen, M. Krisch, D.P. Siddons, and T. Oversluizen. Optical design and performance of the inelastic scattering beamline at the National Synchrotron Light Source. *Rev. Sci. Instrum.* 66 (1995) 1699.
42. Gog, T., G.T. Seidler, D.M. Casa et al. Momentum-resolved resonant and nonresonant inelastic X-ray scattering at the Advanced Photon Source. *Synchrotron Radiat. News* 22 (2009) 12.

43. Hill, J.P., D.S. Coburn, Y.-J. Kim et al. A 2-m inelastic X-ray scattering spectrometer at CMC-XOR, Advanced Photon Source. *J. Synchrotron Radiat.* 14 (2007) 361.
44. Stojanoff, V., K. Hämäläinen, D.P. Siddons et al. A high-resolution X-ray fluorescence spectrometer for near-edge absorption studies. *Rev. Sci. Instrum.* 63 (1992) 1125.
45. Wang, X., M.M. Grush, A.G. Froeschner, and S.P. Cramer. High-resolution X-ray fluorescence and excitation spectroscopy of metalloproteins. *J. Synchrotron Radiat.* 4 (1997) 236.
46. Bergmann, U. and S.P. Cramer. A high-resolution large-acceptance analyzer for X-ray fluorescence and Raman spectroscopy. *Proc. SPIE* 3448 (1998) 198.
47. Glatzel, P., F.M.F. de Groot, and U. Bergmann. Hard X-ray photon-in photon-out spectroscopy. *Synchrotron Radiat. News* 22 (2009) 12.
48. Llorens, I., E. Lahera, W. Delnet et al. High-energy resolution five-crystal spectrometer for high quality fluorescence and absorption measurements on an X-ray absorption spectroscopy beamline. *Rev. Sci. Instrum.* 83 (2012) 063104.
49. Sokaras, D., D. Nordlund, T.C. Wenig et al. A high resolution and large solid angle X-ray Raman spectroscopy end-station at the Stanford Synchrotron Radiation Lightsource. *Rev. Sci. Instrum.* 83 (2012) 043112.
50. Kleymenov, E., J.A. van Bokhoven, C. David et al. Five-element Johann-type X-ray emission spectrometer with a single-photon-counting pixel detector. *Rev. Sci. Instrum.* 82 (2011) 065107.
51. Qian, Q., T.A. Tyson, W.A. Caliebe, and C.C. Kao. High-efficiency high-energy-resolution spectrometer for inelastic X-ray scattering. *J. Phys. Chem. Solids* 66 (2005) 2295.
52. Verbeni, R. M. Kocsis, S. Huotari et al. Advances in crystal analyzers for inelastic X-ray scattering. *J. Phys. Chem. Solids* 66 (2005) 2299.
53. Ishii, L., K. L. Jarrige, M. Yoshida et al. Instrumental upgrades of the RIXS spectrometer at BL11XU at SPring-8. *J. Electron Spectrosc. Rel. Phenom.* 188 (2013) 127.
54. Journel, L., L. El Khoury, T. Marin et al. Performances of a bent-crystal spectrometer adapted to resonant X-ray emission measurements on gas-phase sample. *Rev. Sci. Instrum.* 80 (2009) 093105.
55. Hudson, A.C., W.C. Stolte, D.W. Lindle, and R. Guillemin. Design and performance of a curved-crystal X-ray emission spectrometer. *Rev. Sci. Instrum.* 78 (2007) 053101.
56. Kavčič, M., M. Budnar, A. Mühleisen et al. Design and performance of a versatile curved-crystal spectrometer for high-resolution spectroscopy in the tender X-ray range. *Rev. Sci. Instrum.* 83 (2012) 033113.
57. Pacold, J.I., J.A. Bradley, B.A. Mattern et al. A miniature X-ray emission spectrometer (miniXES) for high-pressure studies in a diamond anvil cell. *J. Synchrotron Radiat.* 18 (2012) 245.
58. Dickinson, B., G.T. Seidler, Z.W. Webb et al. A short working distance multiple crystal X-ray spectrometer. *Rev. Sci. Instrum.* 79 (2008) 123112.
59. Hoszowska, J., J.-Cl. Dousse, J. Kern, and Ch. Rhême. High-resolution von Hamos crystal X-ray spectrometer. *Nucl. Instrum. Meth. Phys. Res. A* 376 (1996) 129.
60. Hoszowska, J. and J.-Cl. Dousse. High-resolution XES and RIXS studies with a von Hamos Bragg crystal spectrometer. *J. Electron Spectrosc. Relat. Phenom.* 137 (2004) 687.
61. Kayser, Y., J. Szlachetko, D. Banas et al. High-energy-resolution grazing emission X-ray fluorescence applied to the characterization of thin Al films on Si. *Spectrochim. Acta B* 88 (2013) 136.
62. Szlachetko, J., J.-Cl. Dousse, J. Hoszowska, M. Berset, W. Cao, M. Szlachetko, and M. Kavčič. Relative detection efficiency of back- and front-illuminated charge-coupled device cameras for X-rays between 1 keV and 18 keV. *Rev. Sci. Instrum.* 78 (2007) 093102.

63. Shevelko, A.P., Y.S. Kasyanov, O.F. Yakushev, and L.V. Knight. Compact focusing von Hamos spectrometer for quantitative X-ray spectroscopy. *Rev. Sci. Instrum.* 73 (2002) 3458.

64. Mattern, B.A., G.T. Seidler, M. Haave et al. A plastic miniature X-ray emission spectrometer based on the cylindrical von Hamos geometry. *Rev. Sci. Instrum.* 83 (2012) 023901.

65. Ice, C.E. and C.J. Sparks Jr. Mosaic crystal X-ray spectrometer to resolve inelastic background from anomalous scattering experiments. *Nucl. Instrum. Meth. Phys.Res.* A 291 (1990) 110.

66. Arkadiev, V.A., A.A. Bjeoumikhov, M. Haschke et al. X-ray analysis with a highly oriented pyrolytic graphite-based von Hamos spectrometer. *Spectrochim. Acta B* 62 (2007) 577.

67. Hayashi, H., M. Kawata, R. Takeda et al. A multi-crystal spectrometer with a two-dimensional position-sensitive detector and contour maps of resonant K beta emission in Mn compound. *J. Electron Spectrosc. Relat. Phenom.* 136 (2004) 191.

68. Hayashi, K., K. Nakajima, K. Fujiwara, and S. Nishikata. Wave-dispersive X-ray spectrometer for simultaneous acquisition of several characteristic lines based on strongly and accurately shaped Ge crystal. *Rev. Sci. Instrum.* 79 (2008) 033110.

69. Alonso-Mori, R., J. Kern, D. Sokaras et al. A multi-crystal wavelength dispersive X-ray spectrometer. *Rev. Sci. Instrum.* 83 (2012) 073114.

70. Kern, J., R. Alonso-Mori, R. Tran et al. Simultaneous femtosecond X-ray spectroscopy and diffraction of photosystem II at room temperature. *Science* 340 (2013) 491.

71. Szlachetko, J., M. Nachtegaal, E. de Boni et al. A von Hamos X-ray spectrometer based on a segmented-type diffraction crystal for single-shot X-ray emission spectroscopy and time-resolved resonant inelastic X-ray scattering studies. *Rev. Sci. Instrum.* 83 (2012) 103105.

72. Szlachetko, J., J. Sá, O.V. Safonova et al. *In situ* hard X-ray quick RIXS to probe dynamic changes in the electronic structure of functional materials. *J. Electron Spectrosc. Rel. Phenom.* 188 (2013) 161.

73. Zachariasen, W. In *Theory of X-Ray Diffraction in Crystals*, Chapter III: Theory of X-Ray Diffraction in Ideal Crystals. New York: J. Wiley & Sons, 1945.

74. Knowles, J.W. A high-resolution flat crystal spectrometer for neutron capture γ-ray studies. *Can. J. Phys.* 37 (1959) 203.

75. Nilsson, S., E. Falkström, and S. Boreving. A flat crystal spectrometer for (n, γ) studies. *Nucl. Instrum. Meth.* 66 (1968) 229.

76. Deslattes, R.D., E.G. Kessler, W.C. Sauder, and A. Henins. Remeasurement of γ-ray reference lines. *Ann. Phys.* (N.Y.) 129 (1980) 378.

77. Kessler, E.G., G.L. Greene, M.S. Dewey, R.D. Deslattes, H.G. Börner, and F. Hoyler. High accuracy, absolute wavelength determination of capture gamma-ray energies for E ~ 5 MeV and the direct determination of binding energies in light nuclei. *J. Phys. G: Nucl. Phys.* 14 (1988) S167.

78. Stritt, N., J. Jolie, M. Jentschel, and H.G. Börner. Study of atomic motion in oriented EuO single crystals using neutrino induced Doppler broadening. *Phys. Rev. Lett.* 78 (1997) 2592.

79. Deslattes, R.D., E.G. Kessler, Jr., L. Jacobs, and W. Schwitz. Selected X-ray data for comparison with theory. *Phys. Lett.* 71A (1979) 411.

80. Kessler, Jr., E.G., R.D. Deslattes, D. Girard, W. Schwitz, L. Jacobs, and O. Renner. Mid-to-high-Z precision x-ray measurements. *Phys. Rev. A* 26 (1982) 2696.

81. DuMond, J.W.M. A high resolving power, curved-crystal focusing spectrometer for short wave-length X-rays and gamma-rays. *Rev. Sci. Instrum.* 18 (1947) 626.

82. Ryde, N. and B. Andersson. A precision curved crystal gamma-ray spectrometer. *Proc. Phys. Soc. B* 68 (1955) 1117.

83. Beckman, O., P. Bergvall, and B. Axelsson. A precision curved crystal X-ray and gamma-ray spectrometer. *Ark. Fysik* 14 (1958) 419.
84. Crowe, K.M. and R.E. Shafer. 7.7 m bent crystal spectrometer at the 184 inch cyclotron. *Rev. Sci. Instrum.* 38 (1967) 1.
85. Piller, O., W. Beer, and J. Kern. Das fokussierende Kristallspektrometer der Universität Fribourg. *Nucl. Instrum. Meth.* 107 (1973) 61.
86. Jett, J.H., N.S.P. King, D.A. Lind, and P. Henning. A diffraction spectrometer for studies of particle excited radiative transitions. *Nucl. Instrum. Meth.* 114 (1974) 301.
87. Borchert, G.L., W. Scheck, and O.W.B. Schult. Curved crystal spectrometer for precise energy measurements of gamma rays from 30 to 1500 keV. *Nucl. Instrum. Meth.* 124 (1975) 107.
88. Koch, H.R., H.G. Börner, J.A. Pinston et al. The curved crystal gamma ray spectrometers "GAMS 1, GAMS 2, GAMS 3" for high resolution (n, γ) measurements at the high flux reactor in Grenoble. *Nucl. Instrum. Meth.* 175 (1980) 401.
89. Borchert, G., J. Bojowald, A. Ercan, H. Labus, T. Rose, and O. Schult. Curved crystal spectrometer for high-resolution in-beam spectroscopy of X-rays and low-energy gamma rays. *Nucl. Instrum. Meth. Phys. Res. A* 245 (1986) 393.
90. Perny, B., J.-Cl. Dousse, M. Gasser et al. DuMond curved crystal spectrometer for in-beam X- and gamma-ray spectroscopy. *Nucl. Instrum. Meth. Phys. Res. A* 267 (1988) 120.
91. Szlachetko, M., M. Berset, J.-Cl. Dousse, J. Hoszowska, and J. Szlachetko. High-resolution Laue-type DuMond curved crystal spectrometer. *Rev. Sci. Instrum.* 84 (2013) 093104.
92. Ludziejewski, T., J. Hoszowska, P. Rymuza et al. In-beam bent-crystal spectrometer for studies of multiple inner shell ionization. *Nucl. Instrum. Meth. Phys. Res. B* 63 (1992) 494.
93. Kern, J., A. Raemy, W. Beer et al. Nuclear levels in $^{192}$Ir. *Nucl. Phys. A* 534 (1991) 77.
94. Mannanal, S., B. Boschung, M. Carlen et al. Nuclear structure of $^{166}$Tm from (α,3n) γ-ray and conversion electron measurements. *Nucl. Phys. A* 582 (1995) 141.
95. Perny, B., J.-Cl. Dousse, M. Gasser et al. High-resolution study of heavy-ion-induced Kα x-ray spectra of molybdenum atoms. *Phys. Rev. A* 36 (1987) 2120.
96. Carlen, M., M. Polasik, B. Boschung et al. M- and L-shell ionization in near-central collisions of 5.5-MeV/amu $^{16}$O ions with Mo atoms deduced from theoretical analysis of high-resolution K x-ray spectra. *Phys. Rev. A* 46 (1992) 3893.
97. Rzadkiewicz, J., D. Chmielewska, Z. Sujkowski et al. High-resolution study of the Kβ₂ x-ray spectra of mid-Z atoms bombarded with 20-MeV/amu $^{12}$C ions. *Phys. Rev. A* 68 (2003) 032713.
98. Eichler, R., B. Aas, W. Beer et al. Energy of a muonic X-ray transition measured with a crystal spectrometer. *Phys. Lett. B* 76 (1978) 231.
99. Beer, W., K. Bos, G.D. Chambrier et al. Crystal spectrometer for measurements of pionic X-rays. *Nucl. Instrum. Meth. Phys. Res. A* 238 (1985) 365.
100. Zhong, Z., D. Chapman, B. Bunker, G. Bunker, R. Fischetti, and C. Segre. A bent Laue analyzer for fluorescence EXAFS detection. *J. Synchrotron Rad.* 6 (1999) 212.
101. Kropf, A.J., R.J. Finch, J.A. Fortner et al. Bent silicon crystal in the Laue geometry to resolve X-ray fluorescence for X-ray absorption spectroscopy. *Rev. Sci. Instrum.* 74 (2003) 4696.
102. Kujala, N.G., C. Karanfil, and R.A. Barrea. High resolution short focal distance Bent Crystal Laue Analyzer for copper K edge X-ray absorption spectroscopy. *Rev. Sci. Instrum.* 82 (2011) 063106.
103. Zhu, D., M. Cammarata, J.M. Feldkamp et al. A single-shot transmissive spectrometer for hard X-ray free electron lasers. *Appl. Phys. Lett.* 101 (2012) 034103.

104. Hiraoka, N., H. Fukui, H. Tanida, H. Toyokawa, Y.Q. Cai, and K.D. Tsuei. An X-ray Raman spectrometer for EXAFS studies on minerals: Bent Laue spectrometer with 20 keV X-rays. *J. Synchrotron Rad.* 20 (2013) 266.
105. Cauchois, Y. Spectrographie des rayons X par transmission d'un faisceau non-canalisé à travers un cristal courbé (I). *J. Phys. Rad.* 3 (1932) 320.
106. Beckmann, O. Photographic bent crystal gamma spectrometer. *Nucl. Instrum.* 3 (1958) 27.
107. Kazi, A.H., N.C. Rasmussen, and H. Mark. Six-meter radius bent-crystal spectrograph for nuclear gamma rays. *Rev. Sci. Instrum.* 31 (1960) 983.
108. Kazi, A.H., N.C. Rasmussen, and H. Mark. Measurement of the Deuteron binding energy using a bent-crystal spectrograph. *Phys. Rev.* 123 (1961) 1310.
109. Beyer, H.F. Characterization of transmission-type curved-crystal X-ray optics for fast ion-beam spectroscopy. *Nucl. Instrum. Meth. Phys. Res. A* 400 (1997) 137.
110. Chatterjee, S., H.F. Beyer, D. Liesen et al. The FOCAL spectrometer for accurate X-ray spectroscopy of fast heavy ions. *Nucl. Instrum. Meth. Phys. Res. B* 245 (2006) 67.
111. Beyer, H.F., D. Attia, D. Banas et al. Crystal optics for hard-X-ray spectroscopy of highly charged ions. *Spectrochim. Acta B* 64 (2009) 736.

# 3 Techniques
## RXES, HR-XAS, HEROS, GIXRF, and GEXRF

*Jakub Szlachetko and Yves Kayser*

## CONTENTS

## 3.1 INTRODUCTION

When X-rays penetrate into matter, a number of photon–atom interactions may occur, the probability of each process being given by its cross section. The cross sections depend on the energy of the X-rays and the specific element under investigation, and their sum is defined as the total absorption coefficient. For elements with $Z > 10$ and in the X-ray energy range from several to few tens of keV (very often defined as the hard X-ray regime), the photoelectric effect is the most dominant process.[1–3] In a single photon bound-electron interaction, the core electron may be excited to higher unoccupied electronic states of an atom, or ejected into the continuum. The probability of each process depends on the energy of incident X-rays as well as on the target element in the sample. Thanks to this correspondence, element specific studies employing hard X-rays as a probe are possible. In the present chapter we focus first on recent developments and experimental results achieved recently with different X-ray spectroscopy techniques and by employing hard X-rays in order to study and determine the electronic structure and hence the chemical state of matter. The use of hard X-rays is a key element. Indeed, because of the penetrating properties of X-rays, the matter may be explored at *in situ* conditions regarding different sample parameters like temperature, gas feed, or pressure.

## 3.2   RESONANT X-RAY EMISSION SPECTROSCOPY

Resonant X-ray emission spectroscopy (RXES), also called *resonant inelastic X-ray scattering* (RIXS) is a widely used tool to probe the electronic structure of matter (there are a number of examples[4–13]). RXES is based on a second-order process of photon–atom interaction in which a core electron is excited into an unoccupied state above the Fermi level by the absorption of an incident X-ray photon of sufficient energy, leading to the creation of an intermediate/excited atomic state. During the decay of the intermediate state, the core hole created in the first step is filled by another inner- or valence-shell electron. The decay from the intermediate to the final state is accompanied by the emission of an X-ray photon; the X-ray energy of the emitted photon depends on electronic levels involved in the scattering process. Thus, the RXES technique relies on monitoring the intensities and energies of the incoming and the emitted X-rays. By tuning the incident X-ray energy around an absorption edge of an element, the unoccupied states are probed by electron excitation to an intermediate state while the information on the occupied electronic levels is obtained by the detection of the specific emitted X-ray photons during an electronic transition leading to the final atomic state. Therefore, RXES allows for determining the overall picture of the electronic states of the investigated sample. We should note here that RXES provides partial information on the density of states because of the selection rules for electron transitions. RXES is dominated mostly by dipole allowed transitions (like $s \rightarrow p$, $p \rightarrow d$), however, in some specific systems (like $3d$ metal-based oxides) quadrupole excitations ($s \rightarrow d$) are also investigated.[14–22]

On a technical basis, RXES requires good energy resolution for both the incident and the detected X-rays. The convolution of these two parameters gives the total experimental resolution, a parameter that is a key component for any RXES experiment. In order to obtain meaningful RXES data, the experimental resolution should be lower than the natural broadening induced by lifetimes of the initial and the final electronic states. The lifetime broadening varies with the involved electronic levels and the measured element. For emitted X-rays with energies between 2 keV and 15 keV, the lifetime broadening of the K-shell is in the range of 0.5 eV–5 eV (i.e., lifetimes of $1.6$–$13 \times 10^{-16}$ s), while for the L-shell it is between 3 eV and 10 eV ($0.7$–$2.1 \times 10^{-16}$ s).[23–25] Due to the small values of the natural lifetime broadening, RXES requires monochromatic X-ray beams and high-energy resolution for X-ray detection. The monochromatic incident X-ray beams are commonly provided at synchrotron sources (see Chapter 1) while the crystal spectrometers (described in Chapter 2) ensure a good energy resolution for the acquisition of the emitted X-rays. Depending on the specific setup used for RXES spectroscopy, an experimental resolution of the order of 0.5–2eV is achieved.[26–32]

Two different types of X-ray spectrometers are employed to measure the RXES signals. The first type is based on a scanning-type geometry (like Johann- or DuMond-type spectrometers, see Chapter 2) where the X-ray emission spectrum is recorded by moving the spectrometer components. At given spectrometer settings, X-rays within a very narrow energy bandwidth are acquired. When changing the X-ray emission energy, both the spectrometer crystals and the detector are moved to a new position for acquisition. The great advantage of such spectrometers is that their peak-to-background ratio allows for the measurement of very weak signals, respectively; experiments with low concentration samples can be realized. However, because of the necessity to scan the X-ray emission energies, the RXES experiments require a certain amount of time for acquision. For this reason, time-resolved RXES studies or experiments with samples sensitive to radiation damage are somehow limited and challenging, all imposed by the time needed to scan the incident and emission X-ray energies.

The second type of spectrometer that is employed in the RXES technique is of the dispersive-type (like Johansson- or von Hamos-type geometries, see Chapter 2) and the X-rays diffracted on the crystal are dispersed onto spatially resolving detectors. The spectrometers can record an X-ray emission spectrum at a fixed position of the spectrometer components, that is, in a single-shot mode. Typically, the energy bandwidth of the X-ray emission spectra is ranging between a few tens to a few hundreds of eV. The dispersive-type spectrometers are characterized by a lower efficiency per eV and a higher peak-to-background ratio. On the other hand, the RXES planes may be acquired at relatively short time scales, limited only by the speed of the monochromator movement for scanning the incident X-ray energy (Figure 3.1). In summary, the best spectrometer setup has to be chosen by taking into account different aspects as well as the goals of the particular experiment, that is, needed time resolution, sample concentration, and sample sensitivity to radiation damage as well as the measured signal strength. Moreover, the different spectrometer setups require distinctive space for operation. Therefore, depending on the sample environment, respectively, the sample cells used for *in situ* studies of some of the experimental setups will have to be excluded because of space requirements in the immediate sample environment.

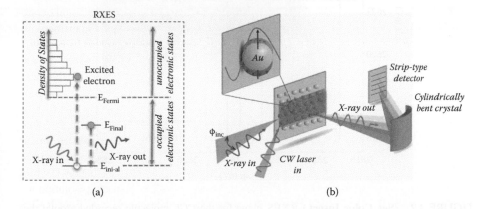

(a)  (b)

**FIGURE 3.1** **(See Color Insert.)** (a) Schematic representation of the resonant X-ray emission process, (b) schematics of an experimental setup for resonant X-ray emission spectroscopy employing laser excitation as the external trigger for plasmon determination. (Reprinted from Sá, J. et al., 2013, *Energy Environ. Sci,.* 6, 3584–3588.[33] With permission.)

### 3.2.1   RXES ON *3p, 4p,* AND *5p* ELEMENTS

RXES spectroscopy in the hard X-ray regime on the p-block of the periodic table covers elements from Al to Ar ($13 < Z < 18$), Ga to Kr ($31 < Z < 36$), and In to Xe ($49 < Z < 54$). The highest partially occupied valence electrons are of *p*-type. Thus, the lowest unoccupied electronic states of the scattering atom can easily be accessed by K-shell spectroscopy (*s* electron), thanks to the strong dipole allowed *s*- to *p*-type excitation. For these elements, energies of incoming and outgoing X-rays are in the range of 1.5 keV to 3.2 keV for the 3rd period and 10 keV to 14 keV for the 4th period, where the elements of both periods can be probed around the K-absorption edge. The elements of the 5th period, In to Xe, can be probed by the L-edge spectroscopy with X-ray energies ranging from 3.7 keV to 6 keV.[34]

As an example, the RXES map recorded around the Cl K-absorption edge for the $CCl_4$ molecule is shown in Figure 3.2.[35] The $\omega_p$ axis corresponds to the incident beam energies and $\omega'_s$ to the energy of emitted photons. The incident beam energy was tuned in the energy range of 2818 eV to 2830 eV in order to excite the K-shell electron of Cl to the lowest unoccupied states of the molecule. The excitation yields were then monitored by detection of the $K\alpha_1$ and $K\alpha_2$ signals, which result from filling the created K-hole by $2p_{3/2}$ and $2p_{1/2}$ electrons, respectively, accompanied by the simultaneous emission of an X-ray photon. The RXES map exhibits

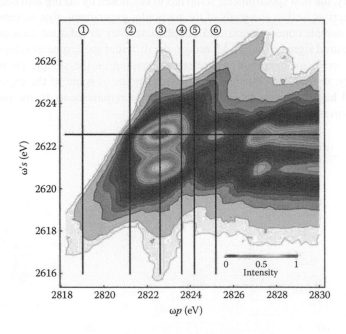

**FIGURE 3.2   (See Color Insert.)** RXES plane for the $CCl_4$ molecule recorded around the Cl K-absorption edge by detecting $K\alpha_1$ ($2p_{3/2} \rightarrow 1s$) and $K\alpha_2$ ($2p_{1/2} \rightarrow 1s$) X-ray emission. (Reprinted from Bohinc, R. et al., 2013, *Journal of Chemical Physics*, 139.[35] With permission.)

rich spectral information. The two resonances observed at incident beam energies of 2822 eV and 2825 eV corresponds to $1s \rightarrow \sigma^*$ excitation, while resonance at 2827 probes $1s \rightarrow 4p$ states. At the highest excitation energies (above 2835 eV) the RXES intensities become almost constant, relating to $1s \rightarrow$ continuum excitation. Note that each of the resonant excitations follows an intensity modulation along a diagonal line in the RXES plane. This diagonal corresponds to a constant energy transfer, that is, constant difference between incident and emitted X-ray energy. Indeed, the resonance excitation processes are driven by the natural lifetime of the core hole and the constant energy transfer intensities are very often used for quantitative and qualitative RXES data analysis. Because of the strong antibonding character of the $\sigma^*$ orbital, the $1s \rightarrow \sigma^*$ excitation results in the dissociation of the molecule.[36–41] Detailed analysis of $1s \rightarrow \sigma^*$ resonances showed breakdown of linear Raman-like dispersion including narrowing of emission lines and asymmetrical peak broadening. The RXES experiments thus allow for the studying of dissociative states and molecular field effects.[42–45] With the help of theoretical simulations, it is possible to evaluate experimental data to confirm the dissociative character of the resonances and to study nuclear dynamics.

Because of the relatively low X-ray energies, such type of experiments have to be performed under vacuum conditions in order to avoid a strong signal absorption in air. This can be performed by placing the X-ray spectrometer setup in a dedicated chamber.[32] However, this also requires a dedicated sample cell, which can sustain a vacuum-ambient pressure difference and simultaneously allows X-rays to penetrate the sample material. Several solutions are possible, most of them based on cells equipped with Kapton or Mylar windows.[32,46,47] Kapton is commonly used in cell designs because of very low X-ray absorption, high resistance to mechanical/pressure forces, and a relatively high melting point.

Because of the high-energy resolution of the incoming and detected X-rays, RXES provides a unique possibility of performing state-selective spectroscopy in order to enhance the chemical sensitivity of the technique.[4,9,10,20,48,49] The difference in the transition probability, resulting from orbital overlaps of different atomic species, provides site-selectivity although the chemical shifts of the core levels are very tiny, or even not detectable. Figure 3.3 shows an RXES plane of a $GaCl_3$ compound recorded around the K-absorption edge of Ga.[50] The incident beam energy was scanned in the energy range of 10368 eV and 10383 eV and the X-ray spectrometer was set to detect X-ray emission resulting from the $3d \rightarrow 1s$ and valence $\rightarrow 1s$ transitions. This type of experiment allows for probing simultaneously the lowest unoccupied and the highest occupied electronic levels in the system.

The RXES data analysis based on the intensity evolution versus the incident beam energy of the $3d \rightarrow 1s$ emission (X-ray energy of 10350 eV) reveals no state selectivity. Indeed, both the $3d$ and $1s$ levels are weakly affected by the chemical surrounding of the Ga atoms, the intensity profiles closely follow the one of XAS recorded in total fluorescence mode. On the other hand, it was found that valence $\rightarrow 1s$ transitions show strong differences, which depend on the chemical state of Ga and, more importantly, are not related to the shifts of the X-ray emission signal. Indeed, the

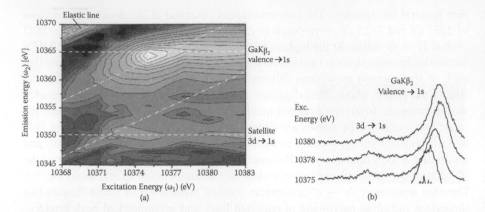

**FIGURE 3.3** RXES of $K\beta_2$ lines of Ga chloride recorded around the K-absorption edge of Ga. (Reprinted from Hayashi, H. et al., 2005, *J. Phys. Chem. Solids,* 66, 2168–2172.[50] With permission.)

orbitals overlap of Ga and the surrounding atoms leads to strong hybridization of valence states that directly affects the decay rates of valence → *1s* emission.[50] Thus, the valence-to-core RXES can provide enhanced sensitivity as compared to core-to-core RXES. However, the data analysis and interpretation have to be supported with theoretical simulations that include electron–electron interaction between excited core- and decaying valence-electrons.[9,10,20,48]

In addition to high chemical sensitivity, RXES can also provide information about ultrafast molecular dynamics by connecting the time and energy information contained in measurements.[35–45] Recently, it was shown that the core hole lifetime of the order of 200 attoseconds set the restrictions to the occurrence of the nuclear dynamics.[11] The corresponding RXES experiment was performed by I $L_3$-edge spectroscopy on $CH_3I$ molecules, and the RXES plane for the $CH_3I$ $L\beta_2$ emission ($N_{4,5}$ → $L_3$ transition) is shown in Figure 3.4. The $L\beta_2$ transition is detected at an emission energy (labeled in Figure 3.4 as scattered photon energy ω) of 4510 eV. The intensity of this emission line varies strongly when tuning the incident beam energy across the $L_3$-absorption edge. The most intense resonance at an excitation energy of 4559 eV corresponds to the transition of $2p_{3/2}$ electron into $15a_1^2$ lowest unoccupied molecular orbital. The Rydberg states are probed by sequential resonance excitations in the incident energy range of 4560 eV to 4564 eV. For energies above higher excitation energies, that is, 4565 eV (marked by an arrow in Figure 3.4), the $L\beta_2$ X-ray emission shows no intensity dependence indicating the $2p_{3/2}$ → continuum excitations. Detailed analysis of RXES data reveals the dispersive character of the resonances, attributed to dissociative molecular states. Because of the very short lifetime of the initial $L_3$ core hole, only tiny relative deviations along constant energy transfer of $2p_{3/2}$ → $15a_1^2$ could be detected. In other words, the time delay between absorption and emission of an X-ray is too short (0.23 femtosecond) in order to probe effectively the energy transfer to the nuclei.[11]

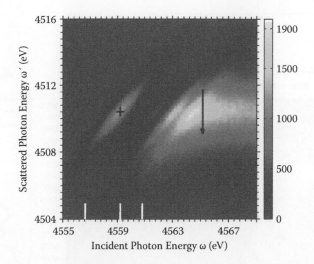

**FIGURE 3.4   (See Color Insert.)** RXES plane of $CH_3I$ measured around the $L_3$-absorption edge of I. (Reprinted from Marchenko, T. et al., 2011, *J. Chem. Phys.* 134, 144308.[11] With permission.)

## 3.2.2   RXES ON *3d*, *4d*, AND *5d* ELEMENTS

The d-block of the periodic table consists of transition metal elements that are characterized by partially filled *d*-orbitals forming the outermost electronic shell. Numerous studies in the hard X-ray regime were carried out to investigate the electronic structure and chemical effects in those elements (there are a number of examples[14–16,20,22,51–55]). The 4th period of the d-block of the periodic table consists of *3d* transition metals, which are accessed by K-shell spectroscopy at X-ray energies ranging from 4500 eV for Sc to 9700 eV for Zn. The RXES spectra of those elements are characterized by unique fingerprints in both X-ray absorption and X-ray emission, allowing for detailed studies of the electronic structure changes of matter under working conditions. The elements of 5th and 6th period (Y to Cd and Hf to Hg, respectively,) are studied by L-edge spectroscopy, where the unoccupied *d*-orbital is probed via dipole-allowed excitations of *p*-core electrons. Here, the X-ray energies that have to be considered in RXES experiments range from 2000 eV to 4000 eV for elements of the 5th period, implying thus, the use of in-vacuum setups. For the 6th period of d-elements, the L-edge excitation requires incident X-rays with energy between 11000 eV and 15000 keV.[34]

Figure 3.5 shows the RXES plane of $TiO_2$ anatase recorded around the K-absorption edge of Ti by detecting the K$\beta$ (*3p* → *1s*) and valence-to-core X-ray emission lines.[56] RXES exhibits multiple information about the electronic states. On one side, the occupied electronic states lying just below the Fermi level may be deduced by non-resonant XES at incident beam energies above 4990 eV (Figure 3.5a). The spectrum consists of strong K$\beta$ X-ray fluorescence at emission energy around 4930 eV, and is accompanied by weak structures lying on the high-energy side. The latter corresponds to the electronic valence orbitals projected onto the *1s* core hole state.

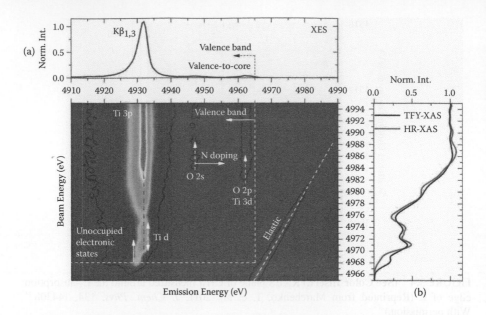

**FIGURE 3.5     (See Color Insert.)** TiO$_2$ anatase RXES plane. (a) Nonresonant XES spectrum; (b) TFY- XAS versus HR-XAS extracted at constant emission energy (4931.7 eV). (Reprinted from Szlachetko, J. and Sá, J. 2013, *Cryst. Eng. Comm.* 15, 2583.[56] With permission.)

It has been demonstrated that these states are due to the O $s$- and $p$-electrons, where O $p$-orbital overlaps with the d-electronic states of Ti. The lowest unoccupied electronic states are probed at excitation energies of 4966 eV to 4976 eV. Those states appear as pre-edge structures at Kβ emission energies. Detailed analysis reveals that those states consist of empty $d$-orbitals of Ti. Moreover, because of the different nature of electron–electron interactions in case of localized and delocalized states, a more detailed picture of Ti d-band could be redrawn from the RXES plane.[19,56,57]

The pre-edge structure is characteristic for all *3d* transition metal oxides. It has been used in many RXES studies as a fingerprint of chemical states and to investigate the electronic structure of metal sites. For example, the pre-edge RXES can be used to study *3d* magnetization in deeply buried thin films (Figure 3.6).[58,59] By working at incident energies around the pre-edge of Fe and measuring Kα X-ray emission (*2p* → *1s* decay), the magnetism can probe both element- and site-selectivity. The experiment reveals that RXES allows for separating the dichroic signal of metal layers from that of ferrite. As a result, in the multilayer systems, RXES may independently probe the magnetization levels of metal and oxide layers.

Site-selectivity is very often explored in RXES experiments.[4,9,10,20,48–50,61] A partial valence-selectivity has been studied, for example, on Co nanoparticles.[62] The RXES data allowed for determining physical properties of atoms on different sites and providing information about the interaction between nanoparticles and nanoparticles' coating. Numerical procedures may be applied to extract pure site-selective evidence including detailed analysis of natural lifetime broadenings included in the RXES spectra. As shown in Figure 3.7, the Kβ (*3p* → *1s*) RXES spectra of Co measured

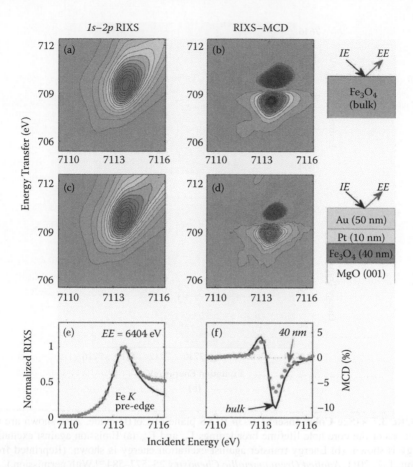

**FIGURE 3.6   (See Color Insert.)** *1s2p* RXES plane (a,c) and its magnetic circular dichroism (b,d) acquired from bulk magnetite (a,b) and deeply buried 40 nm thick film (c,d) performed at H1/43.5 kOe. Constant emission energy (diagonal line plots along the RXES planes) are compared (e,f) in order to show the details of normalization (to the maximum of the pre-edge) and the resulting MCD intensities. The solid line represents the bulk sample; circles represent the buried film. Schemes of the sample structures are shown on the right. Energy transfer (ET) is the difference between incident photon energy (IE) and emitted photon energy (EE) in the RXES process. (Reprinted from Sikora, M. et al., 2012, *J. Appl. Phys.* 111.[58] With permission.)

at the K-absorption edge are affected by the atomic widths of initial (*1s*) and final (*3p*) electronic states. The final state broadens RXES spectra along the X-ray emission axis, while the initial state influences the RXES plane on the diagonal corresponding to constant energy transfer. By transforming the measured RXES data onto the constant energy transfer plane, both contributions are parallel to the energy axes and therefore may be simply accounted in the site-selective analysis of RXES data.[60,63] The initial and final state broadenings in RXES spectra also have another consequence. For closely lying resonances (in the scale of incident and emission X-ray energies), the measured intensities of unoccupied/occupied electronic states

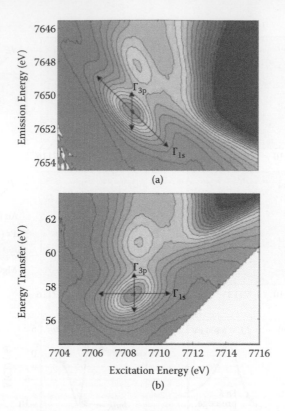

**FIGURE 3.7** **(See Color Insert.)** *1s-3p* RIXS plane of Co(II)-oxide. Also shown are the directions of the core hole lifetime broadenings $\Gamma_{1s}$ and $\Gamma_{3p}$. (a) Emission against excitation energy is shown. (b) Energy transfer against excitation energy is shown. (Reprinted from Kuehn, T.-J., 2011, *Applied Organometallic Chemistry* 25, 577–584.[60] With permission.)

will interfere with each other. This has thus to be considered when interpreting the RXES resonances. One solution to this problem may be the deconvolution of each data point of the RXES plane with respect to both initial and final broadenings.

RXES spectroscopy on *4d* elements around the L-absorption edges is much rarer compared to *3d* transition metal studies. Because of relatively low X-ray energies, the experiments have to be performed in a vacuum environment to avoid strong absorption of X-rays in the air. Such experiments are challenging since the X-ray spectrometer setups for monitoring the X-ray emission signals require a relatively large space for operation and thus the vacuum setups result in relatively large space requirements.[32,64] On the other hand, the L-absorption edge RXES allows probing directly the partially unfilled *4d*-orbital because of the dipole nature of $2p \rightarrow 4d$ excitations (which has been noted in several examples[65–73]). Figure 3.8 presents the RXES plane recorded around the Mo L₃-absorption edge in a $Na_2MoO_4$ compound.[32] The X-ray emission intensities were monitored by monitoring the Lα transition ($3d \rightarrow 2p$). As shown, the RXES plane contains very rich spectral information and reveals a strong contribution of the initial lifetime broadening at constant energy transfer. Thanks to the high-energy resolution of both incoming and detected X-ray

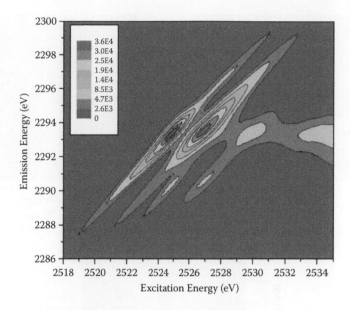

**FIGURE 3.8   (See Color Insert.)** The *2p3d* RXES plane for the tetrahedral coordinated Na2MoO4 system around the Mo L₃-edge. (Reprinted from Kavčič, M. et al., 2012, *Rev. Sci. Instrum.* 83, 033113.[32] With permission.)

energies, the resonances may be identified and analyzed in detail. In the present example, the most intense resonances at incident beam energies between 2522 eV and 2528 eV are due to the ligand field splitting of the *4d* levels in the tetrahedral coordinated $Na_2MoO_4$ system. As a result, RXES overcomes the lifetime broadening, which limits the experimental resolution in typical X-ray absorption experiments and hence allows for the structural characterization.

In the 6th period of *d*-elements, RXES was extensively employed to study the most important elements in catalytic research, namely, Pt, Au, and their oxides (which has been reported in a number of examples[33,54,63,74,75]). The partially unfilled *d*-orbital can be mapped with L-absorption edge spectroscopy to follow electronic structure changes at *in situ* conditions. Spectroscopy on *5d* elements involves relatively high energies of X-rays and thus facilitates any experiments dedicated to investigating matter under working conditions.[76,77] In addition to enhanced sensitivity to probe the electronic and geometric structure of matter, RXES is also a valuable technique to probe the effects of the adsorption of molecules on nanoparticle surfaces. Figure 3.9 shows the L₃-edge RXES of Pt nanoparticles at different working conditions.[74] The RXES map is characterized by a strong $2p \rightarrow 5d$ resonance observed at excitation energies around 11565 eV. The strength and position of this resonance directly reflects the occupancy and chemical state of Pt and may thus be used to monitor electronic changes on the Pt site during reactions or catalytic transformations. Recent experiments on Pt nanoparticles supported on carbon-coated Co-cores reveal changes of CO molecule adsorption geometry in the presence of an external magnetic field. The behavior was related to electronic structure changes on the Pt site induced by a localized magnetic field induced in the Co-cores. The electronic structure change on the

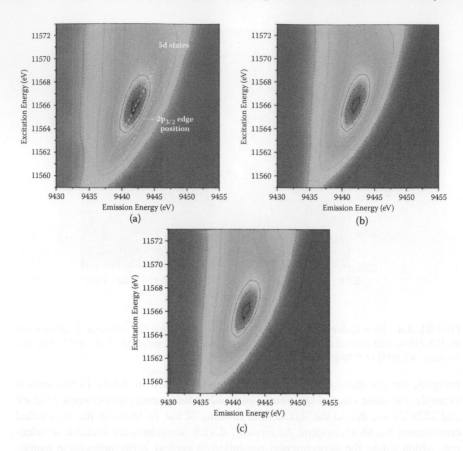

**FIGURE 3.9    (See Color Insert.)** Experimental RXES maps on Pt/Co nanoparticles around $L_3$-absorption edge of Pt. (a) Bare Pt; (b) CO adsorbed on Pt; and (c) CO adsorbed on Pt in the presence of a magnetic field (50 mT). (Reprinted from Sá, J. et al., 2013, *Nanoscale* 5, 8462.[74] With permission.)

Pt site induced simultaneously a change in its catalytic performance.[74] The RXES data analysis in combination with theoretical simulations showed that a part of the CO molecules change adsorption from atop to bridged geometry in the presence of a small magnetic field.

RXES experiments are very often supported by theoretical calculations to obtain deep insight and knowledge of the studied systems. The good agreement between theory and experiment opens the possibility of using theoretical codes to perform preliminary theoretical screening of the metal center electronic structure, that is, to tailor and rationally design novel materials with enhanced catalytic properties. Figure 3.10 shows the RXES plane of IRMOF-3-Si-H complex recorded around the $L_3$-absorption edge of Au by detection of the L$\alpha_1$ X-ray emission signal.[75] Similar to Pt, the RXES map exhibits a resonance at absorption edge relating to $2p \rightarrow 5d$ electronic excitation (diagonal dashed line). The experimental data are evaluated by theoretical calculations involving Slater-type orbitals and the Kramers–Heisenberg

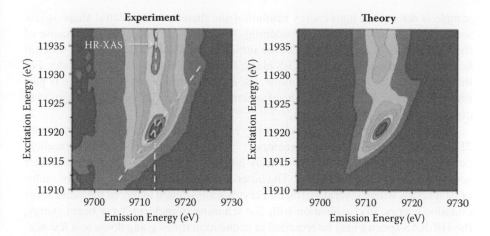

**FIGURE 3.10** **(See Color Insert.)** RXES map of IRMOF-3-Si-H recorded around the $L_3$-absorption edge of Au. The RXES map is compared with the calculated spectrum on the basis of Slater-type orbitals. (Reprinted from Sá, J. et al., 2013, *RSC Adv.* 3, 12043.[75] With permission.)

theorem that describes the RXES process. The combination of experimental and theoretical RXES shows that very tiny modifications on Au electronic structures may be detected by RXES and evaluated in terms of catalytic performances. Such an approach leads to a fine-tune strategy of the electronic structure of gold using postsynthetic modifications. The experimental results reveal that the affected lowest unoccupied molecular orbitals were directly probed by means of RXES technique with extremely high sensitivity.

## 3.3 HIGH-ENERGY RESOLUTION X-RAY ABSORPTION SPECTROSCOPY

The first high-energy resolution X-ray absorption spectrum (HR-XAS), also alternatively called *high-energy resolution fluorescence detected XAS* (HERFD-XAS) was recorded by Eisenberger, Platzman, and Winick in 1976 on the Cu metal K-absorption edge.[78] It has been reported that by employing a high-energy resolution X-ray spectrometer tuned to an emission energy of Cu $K\alpha_1$ narrowing effects are observed in XAS. This approach was then extensively studied in the nineties, first by Hämäläinen, Siddons, Hastings et al.,[79] thanks to the development of both synchrotron sources and X-ray emission spectrometers. Nowadays, the HR-XAS is commonly used for studies of chemical states of matter (of which there are a number of examples[4,8,12,51,80–82]). From a technical point of view, HR-XAS is based on the detection of one X-ray emission energy while scanning the incoming X-ray energy across an absorption edge of an element. Indeed, HR-XAS corresponds to the cut of the RXES plane along the incident X-ray energies where the X-ray emission line shows a maximum intensity. In other words, the HR-XAS technique relies on the same scattering process as described for RXES in Section 3.2. The X-ray emission from the sample is detected at high-energy resolution and therefore the spectral shape is less sensitive to the initial lifetime broadening. As discussed in Section 3.2, because of the nature of the scattering process, the initial electronic state broadens the spectral features at the diagonal of the RXES plane, that is, at a constant energy transfer. For this reason, in the HR-XAS technique, the effect of initial state broadening is partially removed and therefore enhanced spectral features are observed as compared to conventional XAS measurements.[4,8,12,51,80–82]

The most efficient way of performing HR-XAS spectroscopy is realized with 2D focusing Johann-type spectrometers.[26,27] The spectrometers ensure an excellent peak-to-background ratio and a relatively large, as compared to other spectrometer setups, efficiency per energy unit. The latter is further enhanced by the use of multi-crystal setups where several diffraction crystals are aligned to record the same X-ray emission energy. In combination with fast scanning of incident X-ray beam energy, the HR-XAS spectra may be recorded at acquisition times going down to a few seconds (which is noted in some examples[76,83]).

In addition to wide applications in many scientific areas, the HR-XAS technique has to be used carefully. HR-XAS detects only one X-ray emission channel and therefore the recorded intensities are very sensitive to chemical shifts of the X-ray emission lines or to interferences arising from other X-ray emission channels (as first pointed out in Carra, Fabrizio, and Thole[84]). For this reason, the HR-XAS is used instead to detect core-to-core transitions where such interferences are unlikely to happen. Nevertheless, HR-XAS spectroscopy requests from scientists an a priori knowledge or experimental tests to estimate effects of above-mentioned effects. This is of great importance for any *in situ* investigation since the monitored X-ray emission energy for HR-XAS studies has to be set beforehand. If during an *in situ* experiment the X-ray emission line undergoes a chemical shift (as noted in several examples[85–89]), the recorded HR-XAS intensities or HR-XAS edge position may be incorrect. As a rule of thumb, such effects are negligible for higher X-ray energies or X-ray transitions involving deep core levels.

A typical experimental setup for *in situ* HR-XAS investigations is plotted in Figure 3.11. The sample, enclosed in a dedicated for *in situ* study cell, is concealed in a controlled environment. The incoming X-rays impinging on the sample are scanned across an absorption edge of interest. The X-ray emission intensities from the sample is then analyzed by a set of spherically bent crystals aligned to a fixed X-ray emission energy. In the present example, the HR-XAS was used to monitor the oxidation state of the copper phase of a bimetallic catalyst used in the removal of nitrates from water. Figure 3.11b depicts the HR-XAS spectrum of Pt-Cu/Al$_2$O$_3$ after exposure to the reaction mixture.[90] The spectrum was recorded around the K-absorption edge of Cu by setting the spectrometer to the energy of maximum intensity of the Cu K$\alpha_1$ X-ray emission line. The measured *in situ* spectrum is deconvolved using three reference signals, namely, Cu$^{2+}$, Cu$^0$, and Cu alloy. The same approach is used to follow concentration changes and evolution of Cu species by exposing the catalyst to various flows of H$_2$ and N$_2$ in the presence of a solution containing 100 ppm of NO$_3$. It has been found that the concentration of Cu oxidation states depends pronouncedly on the concentration of hydrogen in the feed. Thus,

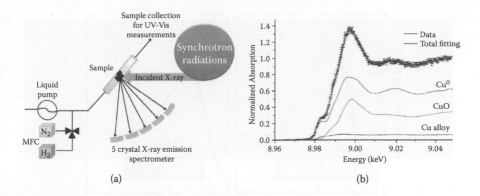

(a)                                        (b)

**FIGURE 3.11** **(See Color Insert.)** (a) Schematic representation of a setup used for *in situ* HR-XAS experiments, (b) linear combination fitting of the normalized HR-XAS spectrum of Pt–Cu/Al$_2$O$_3$ after exposure to 6 mL min$^{-1}$ H$_2$ in the presence of a solution containing 100 ppm of NO$_3$– after 75 min on stream. (Reprinted from Sá, J. et al., 2012, *Catal. Sci. Technol.* 2, 794.[90] With permission.)

similar to the RXES technique, chemical sensitivity of HR-XAS allows for determining the relative population of different metal oxidation states. The experimental HR-XAS data are often supported by theoretical calculations to describe HR-XAS features. Different statistical approaches, like principle component analysis, may be employed to determine the number of chemical species and its concentrations based on *in situ* and reference HR-XAS spectra.

The HR-XAS spectroscopy is also an excellent tool to determine the occupancy of particular electronic states and to follow the electronic structure changes induced by external triggers. As in RXES, the HR-XAS probes the lowest unoccupied states and, in some specific applications, serves as a direct probe of orbitals that are responsible for reactivity. HR-XAS is mostly applied to scientific investigations where subtle changes in the unoccupied electronic states are expected to happen.

Figure 3.12 shows *in situ* HR-XAS spectra of Pt/Rh fuel cell catalyst recorded at the Pt L$_3$-absorption edge.[91] The spectra were recorded applying different electrochemical potentials. The HR-XAS peak recorded at an incident beam energy varying from 11566 eV to 11568 eV, called *whiteline*, corresponds to the $2p_{3/2} \rightarrow 5d$ excitation. The intensity and position of this whiteline feature is therefore a direct measure of Pt $5d$-orbital unoccupied states. Thanks to the application of HR-XAS spectroscopy, the whiteline peak can be detected at enhanced resolution and therefore tiny changes may be revealed allowing for a detailed description of the catalyst under working conditions. The energy resolution aspect is also of prime importance, when the edge position has to be determined with a high precision. Unlike conventional XAS, the HR-XAS technique permits following the edge shifts down to the sub eV regime.[92] The HR-XAS changes depicted in Figure 3.12 relate to the chemisorption of atomic hydrogen- and oxygen-containing species. The data were interpreted based on both edge position and whiteline intensity leading to proof of different molecule adsorption behavior depending on the sample morphology.

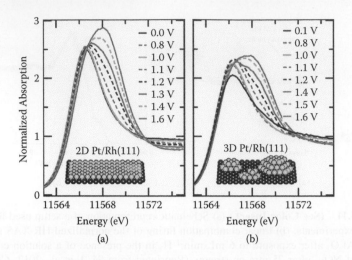

**FIGURE 3.12 (See Color Insert.)** *In situ* Pt L$_3$ HR-XAS for 1 ML Pt/Rh(111) in 0.01 M HClO4: (a) 2D Pt film, (b) 3D Pt islands. Spectra were recorded in the order of increasing electrochemical potentials. (Reprinted from Friebel, D. et al., 2012, *J. Am. Chem. Soc.* 134, 9664–9671.[91] With permission.

## 3.4   HIGH-ENERGY RESOLUTION OFF-RESONANT SPECTROSCOPY

The off-resonant scattering is a second-order process of photon–atom interaction and thus may be regarded as a particular case of RXES. In the off-resonant process, the incoming X-ray energy is tuned well below element absorption. Because of the initial state broadening, the core electron can be nonetheless excited above the Fermi level. This intermediate virtual state then decays radiatively, the initial core hole being filled by another inner- or outer-shell electron with the simultaneous emission of an X-ray photon. The energy of the emitted X-rays depends on the energy of the incident X-rays and the energy of the excited core electron. Therefore, at fixed incident X-ray energies, the energy of the excited electron may be determined by monitoring the emitted X-ray energies. Since the electron excitation yields are proportional to the unoccupied density of states, the latter will be reflected in the intensities of the measured X-ray emission spectra.

Pioneering work in the off-resonant regime was first performed by Sparks in the 1970s employing X-ray tubes and a single crystal as a monochromator.[93] In this work, Sparks studied X-ray emission spectra of different elements at conditions where the incoming X-ray energy is lower than the ionization threshold. Because of low-energy-resolution detectors used in these studies, only asymmetric X-ray emission profiles could be recorded. However, the dependence on the final state energy was clearly observed. In 1982, Tulkki and Åberg developed formulas to describe the RXES process based on the Kramers–Heisenberg formalism.[94,95] Based on their theoretical work, it was concluded that in the off-resonant regime, the X-ray emission spectra exhibit not only an asymmetric line profile, but also the shape of the X-ray emission spectrum is modulated by the unoccupied electronic states of scattering atoms. Since then only few works were performed to investigate the off-resonant regime, mostly due to the low cross sections as compared to RXES. Nevertheless, the influence of the unoccupied electronic states on the shape of the X-ray emission spectra was confirmed experimentally.[96–99] Only recently, by combination of dispersive-type spectrometers and off-resonant excitations, high-energy resolution off-resonant spectroscopy (HEROS) was established.[100,101] The uniqueness of the HEROS approach has a twofold aspect. The use of dispersive-type spectrometers (like von Hamos or Johansson) allows acquiring X-ray emission spectra on a shot-to-shot basis, that is, without scanning components. Since for off-resonant excitations the incident X-ray energy is fixed as well, the entire experiment can be performed with a fixed optical arrangement. Unlike XAS or RXES, where monochromator crystals have to be scanned in order to record a spectrum, HEROS may thus be used to probe the unoccupied electronic states at a physically unlimited time resolution. Indeed, the acquisition time of HEROS is only limited by the efficiency of the experimental setup. The second important aspect of HEROS is the high-energy resolution, which leads to enhanced spectral features. As already noted by Tulkki and Åberg,[94] the X-ray emission spectra in the off-resonant regime are independent on initial lifetime broadening. This was confirmed later by several experiments. To this point, the spectral information of HEROS is indeed the same as that obtained by means of HR-XAS.

The main drawback of HEROS is the experimental efficiency. As compared to the RXES process, the cross sections in the off-resonant regime drop by a factor of $10^2$–$10^3$. Therefore, the application of HEROS narrows down either to concentrated samples or applications at high-intensity pulsed X-ray sources, like X-ray free electron lasers. We should note here that the experimental efficiency could be further improved by using multicrystal arrangements, which allow for experiments with shorter acquisition times or more diluted samples. Nevertheless, HEROS overcomes the major limitation of RXES and HR-XAS—the experimental time resolution. This is a key point for many *in situ* studies where the chemical transformations are not only complex, but also too quick to be followed in real time.

Figure 3.13 shows the principle of HEROS and its correspondence to HR-XAS technique.[100] The X-ray emission spectrum of a Pt foil was recorded at an incident X-ray energy tuned below the $L_3$-absorption edge. The HEROS-XAS spectrum was reconstructed from X-ray emission employing Tulkki and Åberg formulas. As shown, HEROS exhibits enhanced experimental resolution as compared to conventional XAS and indeed the same information as HR-XAS. Because of the fixed optical arrangement, HEROS was applied to *in situ* studies devoted to following the decomposition of platinum acetylacetonate in hydrogen induced by flash heating. Thanks to the subsecond time resolution achieved in this experiment as well as the chemical sensitivity of HEROS, detailed information about the decomposition process was retrieved. The data analysis, based on absorption edge position and whiteline intensity, showed that the decomposition process consists of a two-step process, mediated with a short-lived intermediate.

Another example of HEROS correspondence to HR-XAS measurements is presented in Figure 3.14. The HEROS spectrum was measured around the Xe $L_3$-absorption edge by detecting $3d_{5/2} \rightarrow 2p_{3/2}$ X-ray emission.[101] A detailed analysis of HEROS features showed that the energy positions and oscillator strengths of

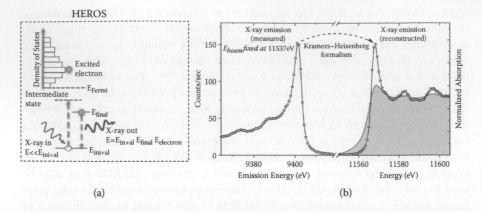

(a)                                              (b)

**FIGURE 3.13    (See Color Insert.)** (a) Schematic representation and energy level diagram for off-resonant scattering process, and (b) the HEROS-XES for the $3d_{5/2}$-$2p_{3/2}$ transition of a Pt foil recorded for a fixed beam excitation energy of 11537 eV. (b) (Blue circles) The reconstructed HEROS-XAS derived from the corresponding HEROS-XES spectrum using the Kramers–Heisenberg formalism. For comparison, the conventional total fluorescence yield spectrum is shown (filled orange area). (Reprinted from Szlachetko, J. et al., 2012, *Chemical Communications* 48, 10898–10900.[100] With permission.)

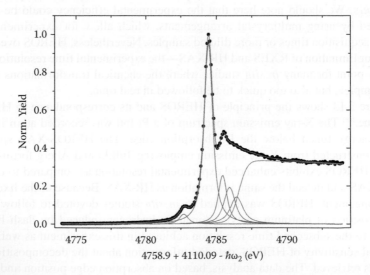

**FIGURE 3.14    (See Color Insert.)** The HEROS $L_3$-absorption spectrum of Xe extracted from the measured $2p_{3/2}$-$3d_{5/2}$ RIXS spectrum, which is recorded at 4758.9 eV excitation energy. The spectrum is fitted with a sum of Voigt functions describing the $[2p_{3/2}]$nd and ns discrete excitations followed by a step function corresponding to the $[2p_{3/2}]$-absorption edge. (Reprinted from Kavčič, M. et al., 2013, *Phys. Rev. B* 87, 075106.[101] With permission.)

discrete excitations may be retrieved and the values are in agreement with previous XAS and RXES experiments. In particular, thanks to an enhanced experimental resolution, being defined by the final state width and incident beam/spectrometer contributions, a weak $2p_{3/2} \rightarrow 6s$ resonance, which is lying just next to a strong $2p_{3/2} \rightarrow 5d$ resonance, can be observed. In other words, this weak structure could not be observed in conventional XAS experiments because of the large broadening of the $L_3$ initial state. As stressed in this work, the uniqueness of the HEROS approach may lead to its application to chemical speciation using either fixed-energy laboratory X-ray sources or monochromatic X-ray pulsed sources.

The chemical speciation by means of HEROS might be performed using the same strategies as for HR-XAS or RXES measurements. First, by measuring a set of reference spectra that can be later used to identify chemical composition of the studied matter. Second, the reference data and experimental spectra may be interpreted based on calculations using common XAS codes. However, to this point, HEROS should be applied in cases where HR-XAS and RXES techniques are limited by the experimental time needed for the acquisition.

## 3.5 SURFACE-SENSITIVE XES

While XES allows for the chemical speciation of a sample by probing the occupied density of states (DOS), the analysis of possible catalysis applications calls for an enhancement of the surface-sensitivity for the following reasons. On the one hand, it is necessary to distinguish whether photons are emitted by atoms localized close to the surface or by atoms localized in the bulk volume; on the other hand, only the sample region that is affected in the first place by a chemical process should be subjected to the measurement. As will be shown, setups based on a grazing incidence or grazing emission geometry with respect to the incident, respectively, emitted X-rays allow for doing so. The refraction of the X-rays at smooth, flat, and sharp sample interfaces, that is, at the border between two media with different optical properties, has to be considered to explain the surface-sensitive character. When performed with a high-energy resolution setup, experiments realized in a grazing geometry allow, in addition to the information contained in the XES signal, gathering information on the spatial distribution of the studied element. For example, the XES intensity dependence on the grazing angle is pronouncedly different for bulk-, layer-, and particle-like distributions. Information on the density, the interface roughness and the elemental and sometimes chemical composition can be retrieved as well. Thus, the grazing incidence and the grazing emission geometry offer complementary information on the surface-near region or the studied chemical system deposited on the top of a suitable surface.

### 3.5.1 PROBING THE SURFACE WITH X-RAYS

To probe the surface and the near-surface region of a sample, a variety of methods can be chosen from.[102] Among them one can name, for example, secondary ion mass spectroscopy (SIMS),[103] respectively, time-of-flight (TOF) SIMS,[104] Rutherford backscattering (RBS),[105] medium energy ion scattering (MEIS),[106] techniques based on electrons[107,108] like Auger electron spectroscopy (AES) or X-ray photoelectron spectroscopy (XPS), ellipsometry,[109] and scanning electron microscopy (SEM)[110] or scanning tunneling microscopy (STM).[111] Each method has its inherent limitations in terms of sample consumption, cost of analysis, required sample environment, quantification, chemical and elemental sensitivity, lateral and depth resolution, and accessible depth region. Often a combination of different techniques is necessary to gather the required information. Examples of comparative studies of different analytical techniques can be found in the literature (as noted in some examples[112,113]).

With respect to other surface-sensitive analysis techniques, X-ray based approaches like X-ray diffraction, X-ray scattering, XAS or XES are competitive in terms of surface characterization.[114] Moreover, surface studies by means of X-rays are non-consumptive and do not necessarily require a high-vacuum environment like particle-based methods. At the same time, X-ray based techniques require little or no sample preparation, can be applied to a wide range of materials regarding the elemental and chemical composition as well as the concentration, allow for quantitative measurements, and are characterized by a high surface-sensitivity if combined with a grazing incidence or emission geometry. Indeed, in standard X-ray probe based XES set-ups the soft interaction, compared to (charged) particles, of hard X-rays with matter and the inherent rather large penetration depth of the incident primary X-rays into the sample results usually in a reduced efficiency for surface-sensitive studies. Indeed, the recorded XES signal emanates not only from the surface but also from the bulk volume of the probed sample. The large, micrometer-range penetration of X-rays has to be taken into account to explain the limited efficiency for surface analysis by means of X-ray probing. The XES signal produced in the vicinity of the surface may, indeed, be hidden in the XES signal from the bulk or in the background (due to elastic scattering, detector noise, and photoelectron Bremsstrahlung) if it is not located in a clean, that is, free of any other signals, region of the acquired X-ray spectrum.

In grazing incidence or grazing emission conditions, the depth (measured perpendicularly from the sample surface into the bulk) of the sample volume, which contributes to the experiment, varies from a few nanometers at the smallest grazing angles to several hundred nanometers at larger grazing angles. The grazing angle is defined on the outer surface of the sample as the angle subtended between the X-ray path of interest (either the one of the incident X-rays in the case of grazing incidence geometry or the one of the emitted X-rays in the case of grazing emission geometry) and the interface formed by the sample surface with the surrounding environment (air, gas, or vacuum). Because of the wide range of accessible depth regions, there are fewer restrictions regarding the sample thickness that can be analyzed compared to SIMS, AES, or XPS. Indeed, in AES and XPS, which offer a good depth resolution as well as elemental and chemical sensitivity but strive to deliver quantitative results, the escape depth of the electrons limits the depth region below the surface which can be analyzed. The electrons typically have an energy of a few keV or even less, thus, the mean free path is at the most a few tens of nanometers. Sputter-based techniques such as SIMS imply not only sample consumption. The first nanometers below the surface in addition cannot be properly characterized because of the transient region defined as the depth region in which the sputter yield cannot yet be assumed to be constant. Furthermore, the sputtering rate depends on the elemental and chemical sample composition. Thus, for a reliable calibration of the depth scale, which correlates with the sputter rate, reference samples are often required in sputter-based approaches.

Alternatively, particles could be used to excite the XES signal in the near-surface region, like in particle induced X-ray emission (PIXE),[115] since the penetration depths of particles into the sample are much shorter compared to hard X-ray photons, although not necessarily confined to the first few nanometers below the sample surface. Moreover, particle induced XES requires a high-vacuum setup to avoid the absorption of the particles in air or gas and results in noisier X-ray spectra. Indeed, particles have compared to X-ray photons a larger probability to multiply ionize atoms, which makes the quantitative interpretation of the spectra more difficult due to the presence of many satellite lines. In addition, a quite noisy background radiation is induced due to the particle Bremsstrahlung produced in the bulk target. XES spectra induced by photons, on the other hand, are much cleaner. This proves to be especially useful in cases where the element to be analyzed is only present at low concentrations. Other advantages of hard X-ray induced XES are a nonconsumptive analysis and little requirement on the sample preparation for the analysis itself. The dynamic range and the linear response of X-ray detectors are also advantageous, especially in quantitative measurements.[116] Furthermore, in comparison to other analytical techniques, an operation under atmospheric pressure can be envisaged with hard X-rays, allowing for *in situ* experiments. The latter aspect is especially of importance if chemical reactions, respectively, the chemical state of one or more reactants before, during, and after the reaction are to be surveyed.

As it has been mentioned, X-ray probe based XES can be rendered surface-sensitive by either reducing the incidence angle of the probing X-ray beam to grazing incidence angles or by recording only the part of the XES signal which is emitted at a grazing emission angle (Figure 3.15). In the first approach, called *grazing incidence*

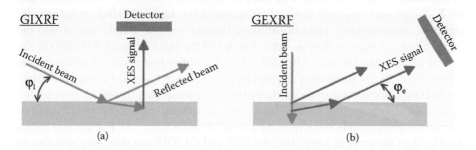

(a)                                                                                    (b)

**FIGURE 3.15** Basic scheme of a GIXRF (a) and of a GEXRF setup (b). In GIXRF, the sample surface is irradiated by a (monochromatic) X-ray beam at a well-defined shallow incidence angle (which may be tuned to different values) while the emitted XES signal is detected at large emission angles. In GEXRF, an X-ray beam, which is incident at macroscopic incidence angles, excites the XES signal, which is detected at one or several well-defined shallow emission angles. Note that because of the difference in the refractive index the angle subtended between the X-ray path and the refracting interface formed by the sample and the surrounding environment is smaller on the sample side.

*X-ray fluorescence* (GIXRF), the volume in which the XES signal is excited is confined to the surface-near region, in the second one, called *grazing emission X-ray fluorescence* (GEXRF), only the XES signal emitted by surface-near atoms is detected and the XES signal originating from atoms in the bulk is suppressed. The basic principle of both grazing XRF geometries consists, indeed, in enhancing the sensitivity toward the XES signal emitted by surface-near atoms with respect to the XES signal emitted by atoms localized in the bulk volume of the sample.

Two factors have to be taken into account to explain the surface-sensitive character of an XES measurement at grazing incidence or emission angles, one being the refraction of the X-rays at the interface formed by the sample and the surrounding environment, the other one being the increased absorption in the depth direction of the incident X-ray beam, respectively, the emitted XES signal, caused by the implicit long incidence, respectively, emission paths within the sample. The refraction factor is of importance since for energies in the X-ray domain, solid samples are optically less dense than the surrounding environment. This can be explained as follows.[117,118] In the derivation of the refractive index $n$ in the Lorentz theory, the electrons are assumed to be quasi-elastically bound and forced to oscillations by incident X-rays. The frequency of the latter is of the order of the binding frequency of the atomic electrons. Upon this, the electrons radiate with a phase difference and by superposition of both radiations, the phase velocity $v$ of the primary beam is altered to values larger than $c$, the speed of light. Hence, the refractive index $n = c/v$ is modified to values smaller than one by a decrement factor $\delta$. This is not in contradiction with the relativity theory, since the group velocity, that is, the speed at which the signal is transported, does not exceed $c$.

### 3.5.2 GRAZING INCIDENCE SPECTROSCOPY

Since in the X-ray energy range the real part of the complex refractive index (related to the scattering properties) of solid samples is smaller than one by a decrement factor $\delta$, X-rays undergo total external reflection at the sample surface at sufficiently small incidence angles. The critical angle depends on the sample matrix and the X-ray energy of interest, that is, either the one of the incident beam (GIXRF) or the one of the emitted photons (GEXRF). A.H. Compton first reported total external reflection of X-rays from solid samples in 1923,[119] whereas L.G. Parratt established the first theoretical formalism on the basis of the dispersion theory in 1954 to discuss surface properties of solids.[120] First applications, oriented toward the tracing of minute sample amounts, were presented about 20 years later.[121,122] For grazing angles smaller than the critical angle, both GIXRF and GEXRF are thus only sensitive to the surface itself and the first few nanometers below it. A contribution from the bulk can be excluded since either no XES signal is excited in the bulk volume (GIXRF) or the XES signal emitted from the bulk volume cannot be detected because of the refraction at the interface formed by the sample and the surrounding environment (GEXRF). A requirement for experiments at grazing incidence or grazing emission are refracting interfaces which are optically flat and smooth, thus characterized by a very low roughness and the absence of waviness within the irradiated area, and which present a sharp (compared to the X-ray wavelength) optical boundary. Otherwise, the grazing incidence and emission angle, respectively, cannot be properly defined or be set to values below the critical angle. The effects of rough surfaces or layer interfaces in the grazing incidence geometry are widely discussed.[123-126]

For angles above the critical angle of total external reflection, X-ray absorption limits the depth of the sample volume contributing to the measurement, and the measurement is characterized by a somewhat increased sensitivity to the composition of the sample matrix. In a first approximation, the probed depth increases in this angular regime linearly with the grazing angle.

The conditions for total external reflection of X-rays can be derived from Snell's law (also known as *Snell–Descartes law*). The latter can be deduced from the continuity condition for the incident, reflected and refracted electromagnetic waves at the interface requiring that the temporal and spatial evolution of the three waves shall be identical at the interface

$$\cos(\varphi_i) = n * \cos(\varphi_t),  \tag{3.1}$$

where $\varphi_i$ represents the incidence angle on the sample, $\varphi_t$ the angle with respect to the interface after transmission through the latter, and the complex refractive index $n$ can be written as

$$n = 1 - \delta + I * \beta,  \tag{3.2}$$

$\delta$ being the refractive index decrement and $\beta$ the absorption index. Both are positive quantities with values in the range of $10^{-3}$–$10^{-7}$ and are either related to the scattering or to the absorption properties,[117]

$$\delta = \frac{N_A}{2*\pi} * r_e * \rho/A * f(\lambda_i) * \lambda_i^2,  \tag{3.3}$$

$$\beta = \frac{\mu(\lambda_i)}{4*\pi} * \rho * \lambda_i.  \tag{3.4}$$

The wavelength of the incident X-rays is represented by $\lambda_i$, $N_A = 6.022 * 10^{23}$ corresponds to the Avogadro's number, $r_e = 2.818 * 10^{-3}$ Å to the electron radius (or equivalently the X-ray scattering amplitude per electron), $\rho$ to the mass density of the scattering sample, $A$ to the molar mass of the sample, $\mu(\lambda_i)$ to the total mass absorption coefficient, and $f = f(0) + f_1(\lambda_i)$ to the real part of the atomic scattering factor, where $f(0)$ stands for the atomic forward scattering factor, which is approximately equal to the atomic number $Z$ of the sample's element (the difference being given by a small relativistic correction[127]) and $f_1(\lambda_i)$ for a correction factor that is essential in the X-ray wavelength domain below the absorption edges of the sample's element. In the formulas above, a homogeneous mono-elemental sample was assumed but the formulas can also be applied to compound materials if the scattering factor $f$ is known for the considered compound. All other factors can be calculated with a weighted linear combination of the corresponding elemental factors. The weight coefficients are given by the relative elemental concentration. The ratio $N_A \rho/A$ corresponds to the atomic density, which can be related with $f(0)$ to the electronic density, the electrons being the X-ray scatterers.

The critical angle for total external reflection can be approximated by

$$\varphi_c \approx \sqrt{2*\delta} \approx \frac{\lambda_i}{\sqrt{\pi}} * \sqrt{\frac{N_A * r_e * \rho * Z}{A}}. \tag{3.5}$$

The latter approximation is valid for X-ray wavelengths shorter than the wavelengths corresponding to the absorption edges of the sample's element. The critical angle depends on the sample's elemental composition and the wavelength of the incident X-rays and tends to decrease for heavier elements and higher X-ray energies.

However, for grazing incidence angles equal to or smaller than the critical angle for total external reflection, the incident X-ray photons are not only totally reflected at the refracting interface formed by the sample and its environment. There exists also an evanescent, exponentially damped wave, propagating along the surface and thus penetrating into the sample. The short, vertical penetration range is due to energy and momentum conservation. The angle subtended by the reflected beam with respect to the refracting interface is identical to the incidence angle because of the continuity condition at the interface. The incident and reflected beam will superpose coherently, creating a standing X-ray wave pattern.

The total external reflection of X-rays not only improves the excitation efficiency for fluorescence radiation of the near-surface region but in addition, prevents any fluorescence excitation of the bulk (see Figure 3.16). For incidence angles larger than the critical angle for total external reflection the incident X-ray radiation penetrates into the sample, but due to the shallow incidence angles, the X-ray absorption is quite pronounced in the depth direction (factor $\sin \varphi_i$), limiting the depth region, which is effectively excited.

The standing X-ray wave pattern above the reflecting surface results from the coherent superposition of the incident and reflected X-ray beams, and is characterized by regions with constructive and destructive interference, respectively. For a plane, monochromatic wave (amplitude $E_0$, wave vector $\vec{k}$, wavelength $\lambda_i$) incident at an angle $\varphi_i$ relative to the surface and the corresponding plane wave reflected by the surface, the standing wave field is described by

$$\vec{E}(\vec{r},t) = \vec{E}_0 * \cos\left( k * \sin(\varphi_i) * z + \frac{\Delta\Psi_R}{2} \right) * \exp\left( i \left( \varpi * t + k * \cos(\varphi_i) * x + \frac{\Delta\Psi_R}{2} \right) \right), \tag{3.6}$$

where $k = 2\pi/\lambda_i$ is the norm of the wave vector $\vec{k}$ and $\Psi_R$ is the phase difference between the incident and the reflected plane X-ray wave due to scattering. Moreover, a surface reflectivity equal to 1 has been assumed. The coordinate system has been fixed such that the $xy$-plane corresponds to the surface, and the $z$-axis is perpendicular

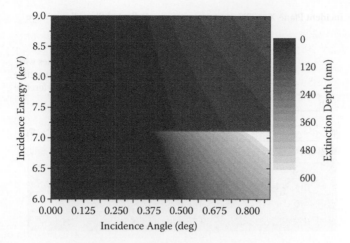

**FIGURE 3.16** Plot of the extinction depth as a function of the incidence angle and the incident beam energy for a Fe sample. The extinction depth corresponds to the perpendicular distance from the surface after which the intensity has been attenuated by a factor $e^{-1}$. The extinction depth depends, at a fixed incident beam energy, strongly on the incidence angle. Below the critical angle (marked by a steep step in the variation of the extinction depth) only a shallow surface layer (about 3–5 nm) is penetrated, whereas for larger incidence angles the incident X-ray photons penetrate deeper into the sample. The extinction depth varies inversely with incident X-ray energy, except for energies around the K-edge of Fe (7.112 keV) where the extinction depth varies sharply. This is of importance for absorption measurements. For comparison, in setups where the probing X-ray beam is incident perpendicularly to the surface, the attenuation length is at least 3.1 µm for the displayed beam energies.

to the surface, pointing out from it, and the origin of the axis coincides with the sample surface (the $xy$-plane). Furthermore, the incident wave vector is confined in the $xz$-plane and the spatial and temporal evolution of the incident wave vector and of the sum of the refracted and transmitted wave vectors are required to be identical. The theoretical description for the standing wave pattern due to the coherent superposition of the incident and diffracted beams is based on the dynamical diffraction theory.

From the equation of the X-ray standing wave field one can deduce that the latter presents at different, regular intervals above the reflecting surface planes with either constructive or destructive interference, that is, zero or maximum amplitude, which is parallel to the surface (Figure 3.17). The periodicity of these planes is equal to $\lambda_i/$ $(2 \sin \varphi_i)$. Since the periodicity is a function of the incidence angle, changing the latter allows altering the created standing wave field. The period of the standing wave pattern will vary from infinity at 0° incidence angle to

$$D_{crit} = \frac{\lambda_i}{2 * \sin(\varphi_c)} \approx \frac{\lambda_i}{2 * \varphi_c} \approx \frac{\lambda_i}{2 * \sqrt{2 * \delta}} = \frac{\sqrt{\pi}}{2} * \sqrt{\frac{A}{N_A * r_e * \rho * Z}} \quad (3.7)$$

at the critical angle. Note that the latter value is independent of the incident X-ray wavelength and depends solely on the reflecting sample.

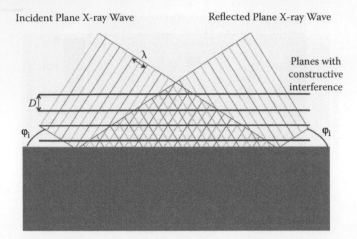

**FIGURE 3.17** X-ray standing wave field created by the interference of an incident and a totally reflected X-ray plane wave of wavelength $\lambda$. The maximum and minimum amplitudes of the plane waves are displayed in light and dark gray, respectively. The planes with constructive or destructive interference are parallel to the sample surface. The periodicity, which can also be deduced from geometric considerations, varies with the incidence angle and the incident wavelength.

The boundary condition at the reflecting interface implies that in the case of total reflection, the phase shift between the incident wave and the reflected wave varies from $\pi$ to 0 if the incidence angle varies from 0° to the critical angle. Thus, at 0° incidence, a plane with destructive interference is coincident with the surface while at the critical angle a plane with constructive interference is coincident with the reflecting interface. In the latter case, the amplitude of the total wave vector (the sum of the incident and the reflected wave vector) is doubled and the intensity is multiplied by four (assuming a reflectivity $R$ of 100%). The intensity in the general case where the reflectivity is not one is given in Bedzyk, Bommarito, and Schildkraut.[128]

$$I(\varphi_i, z) = \left| E_0^2 \right| * \left( 1 + R + 2 * \sqrt{R} * \cos\left( \arccos\left( \frac{2 * \varphi_i^2}{\varphi_c^2} - 1 \right) - \frac{4 * \pi * z * \sin(\varphi_i)}{\lambda_i} \right) \right).$$

(3.8)

The maximum and minimum values of $I(\varphi_i, z)$ define the antinodal and nodal lines, that is, the planes with destructive and constructive interference, respectively. Since the intensity varies continuously,[16] the intensity at the surface ($z = 0$) is also given by the above equation. Below the surface the intensity is exponentially damped due to X-ray absorption (Figure 3.16).

The surface region, over which the standing wave pattern extends, varies inversely with the incidence angle. The X-ray source coherence lengths have also a major impact on the extension of the standing wave pattern, especially in the height direction[129,130] (along the $z$-axis). The contrast of the antinodal and nodal lines can fade away.[131]

Because of the interference of the incident and reflected wave, the probability of exciting the XES signal is increased in the volume covered by the standing wave field. This is, for example, of importance for particles deposited on a reflecting surface. An X-ray standing wave field can also be created within a layered sample if the refractive index of the underlying substrate is smaller than the one of the covering layer, that is, for the incident photon energy, the critical angle for the bulk material is larger than that of the layer material. This leads to an enhanced excitation efficiency of the layer atoms while in general the critical angle for the layer allows deducing the density of the surface layer. In the literature descriptions can be found of algorithms to calculate the incident angle dependent X-ray standing wave field for different sample types[132,133] (from which the angle-dependent excited XES intensity can be extracted) and of simulation software of the standing wave field and the XES angular intensity dependence.[134,135]

It should be noted that one distinguishes between an X-ray standing wave field created under total reflection conditions and an X-ray standing wave field created by Bragg diffraction. In the latter case, the standing wave field results from the superposition of an incident coherent and monochromatic X-ray beam with the Bragg diffracted beam, that is, the coherently scattered beam from crystal planes or the interfaces in a multilayer. B.W. Batterman, who used it to trace interstitial impurities in crystalline samples, has pioneered this approach.[136] With respect to the total reflection case, the incidence angles are larger since the angular region around the Bragg angle is scanned, the Bragg angle $\theta_m$ being given by the Bragg law

$$m * \lambda = 2d * \sin(\theta_m) \qquad (3.9)$$

where d is the lattice spacing of the diffracting planes and m the diffraction order. Thus, the beam footprint on the sample surface is smaller. The standing wave field is formed both above and below the sample surface. Since the periodicity of the wave field is given by the lattice spacing of the diffracting sample, it is much shorter than in the total reflection case. The planes with constructive and destructive interference are furthermore parallel to the planes on which the X-rays are diffracted (crystal planes or multilayer interfaces). Further differences in the Bragg diffraction case are a lower peak reflectivity and more complex conditions to fix the phase of the diffracted beam.[137] An angular scan through the Bragg region, that is, the angular range in the vicinity of the Bragg angle, corresponds to a phase shift of $\pi$ or equivalently to a shift of the nodal and antinodal lines by half of the lattice spacing d. For the smallest angle the nodal lines are on the diffracting planes, and for the largest angle, the antinodal lines. Outside the Bragg region, the modulation of the wave field is lost as the intensity of the diffracted beam decreases strongly.[138]

Regarding the setup required for an experiment under total reflection conditions it can be said that in general it corresponds to a special configuration of an energy-dispersive X-ray fluorescence (EDXRF) setup, the main difference being that a standing wave field whose intensity oscillates between 0 and up to fourfold the incident intensity is created on the top of the sample surface for sufficiently small incidence angles.[139] Thus, in comparison to EDXRF, the excitation of the XES signal of particles above the surface is considerably enhanced since the probability for photoelectric absorption is correspondingly locally enhanced while at the same time background contributions from the bulk volume are suppressed. As a consequence a setup based on the total external reflection of X-rays presents a drastically improved signal-to-background ratio compared to an EDXRF setup. An energy-dispersive detector, allowing for simultaneous multielemental detection, mounted close to the sample surface in order to increase the solid angle of detection and to thus ensure an efficient XES signal detection, records the excited XES signal.

The primary X-ray beam used to excite the XES beam needs to provide sufficient flux on the sample to realize the measurements in a reasonable time interval. In the laboratory a high power, fine focus X-ray tube has to be used; alternatively, the experiment can be conducted at a synchrotron radiation beam line. In the latter case, energy-tunability, monochromaticity, linear polarization, and low beam divergence can be profited from. The linear polarization allows for geometrical arrangements with reduced background intensity from scattering events.[140] A low beam divergence is of importance for the angular resolution of the incidence angle. If the incident beam has a large energy bandwidth, a low-pass filter is often used to allow for larger incidence angles while still fulfilling the condition for total external reflection. Indeed, the critical angle depends inversely on the photon energy. Different setup configurations and approaches to produce a (monochromatic) low-divergence beam have been mentioned.[141–144] Depending on the purpose of a measurement at grazing incidence conditions, two configurations can be distinguished.

If the aim of the measurement is to experimentally assess the quantitative elemental or chemical composition of a deposit on the top of a reflecting surface or of a thin layer on the top of a bulk sample, measurements of the XES signal at a single incidence angle below the critical angle for the largest incident X-ray photon energy will be sufficient. This approach is usually referred to as the *total reflection X-ray fluorescence* (TXRF) technique.[117,130,141,145–147] The incidence angle may be adjusted by pressing the sample against a reference holder. For a monochromatic incident X-ray beam, an incidence angle of about 71% of the critical angle is usually chosen. This incidence angle is called the *isoradiant angle*.[148] For identical elemental concentrations, the detected XES intensity will be independent of the particle distribution (particle-like or layer-like surface morphology). Quantification can be realized by adding an internal reference standard, by a reference-free approach based on an exact calibration of the instrumentation and the knowledge of the atomic fundamental parameters[149] or by solely theoretical calculations.[150] Reference samples can be used as well,[151] absorption of the incident and emitted radiation in the quantification of an unknown sample can be neglected up to the critical thickness.[152]

The micro and trace analysis capabilities offered by the TXRF technique[153] have among others been profited from in environmental sciences,[154] pharmacological applications,[155] biology,[156] petrochemistry,[157] and surface contamination control on semiconductor samples.[140,158,159]

If the goal is to access, in addition to the elemental or chemical information, spatial information about the different sources emitting the XES signal, an angle-resolved measurement is required. The corresponding techniques are called *angle-resolved*

*TXRF, GIXRF,* or *X-ray standing wave (XSW) techniques under grazing incidence conditions.* The intensity of the XES signal is monitored as a function of the incidence angle, the grazing angle being varied from 0° incidence to typically twice or even thrice the critical angle. The evolution of the XES signal with the grazing incidence angle permits not only to probe different sample depths but also to distinguish between different types of surface depositions or types of samples: bulk samples, (buried) layers, multilayers, implanted samples, and residual grains. For example, grains, particles, or residues on the top of a reflecting surface are the most efficiently excited for incidence angles below the critical angle while (buried) layers and implanted atoms are the most efficiently excited for incidence angles in the vicinity of the critical angle. Thus, particle- and layer-like surface structures can be distinguished qualitatively by simply comparing the XES intensity of a sample at two different incidence angles below the critical angle.[160] Above the critical angle, the XES intensity for these sample types is roughly constant while bulk samples are characterized by an increasing XES intensity for increasing incidence angles. Lateral information can be acquired by scanning techniques.[161]

For angle-resolved grazing incidence measurements, a monochromatic incident beam is required in order to have a well-defined critical angle and to avoid a blurring of the correlation between the standing wave pattern and the local distribution of the atoms emitting the recorded XES signal. The sample needs to be placed on an adequate positioning system since the incidence angle needs to be known precisely and varied accurately. Geometrical effects due to the varying incidence angle (and the consequently varying footprint of the incident beam on the sample surface) have been discussed in many examples.[149,162,163]

Applications realized by means of the GIXRF or XSW technique under grazing incidence conditions aim, in addition to the elemental and/or chemical characterization, at performing depth-resolved studies or at the extraction of structural information, that is, the (depth) distribution of the atoms emitting the XES signal. A few examples will be mentioned in the following.

The thickness and sequence of buried nanolayers composed of low-Z elements was investigated by recording the dependence of the XES intensity on the grazing incidence angle.[164] While the extracted thicknesses were compared to results obtained with standard XRF, the GIXRF measurements, which were performed at low X-ray energies, permitted moreover to reveal a carbon contamination at the surface. The thickness, the density, and also the roughness of the different layers in multilayered samples were assessed as well in different studies.[165,166] The compositional depth profile of thin films used for solar cells was also characterized by means of GIXRF.[167] Studies of the depth-dependent elemental contamination of the vertical sidewalls of structured Si wafers[168] and the surface oxidation of Cu layers after evaporation on Si wafers have also been reported.[169] Single layers have been characterized in terms of thickness on a nanometer scale,[133,170,171] the thickness and the density of the films as well as the elemental amount or quantity of deposited material were assessed. The assumption of a stratified model with homogeneous and discrete layers was made to fit the data by inverse modeling, the initial model needing to describe the physical sample already well enough for a reliable interpretation.[133,169]

Furthermore, molecules in a liquid film have been investigated,[172] as well as the diffusion of Au particles in a polymer film[173] and the ionic concentration in an electrolyte solution during titration.[174] In another application, the diffusion of Cu and Au into the Si wafer during annealing was studied.[175]

Furthermore, nanosized particles deposited on a flat, reflecting surface were probed quantitatively.[176] The geometrical arrangement and an eventual total reflection on the particle surfaces had to be taken into account. The average vertical size of dispersed metallic nanoparticles on an Si or a multilayer structure has been evaluated as well.[177] A few limitations (particle size, finite coherence of the wave field) for the characterization of particles have, however, to be considered[178] as well as the surface coverage with particles. Implanted solar wind particles were analyzed too, the sensitivity of the technique proving to be useful to distinguish the ions collected during a NASA mission from the terrestrial surface contamination caused by a crash during landing.[179]

The depth distribution of dopants implanted in semiconductor wafers has been assessed by means of GIXRF[180] and GIXRF combined with results obtained from SIMS measurements.[181]

### 3.5.3 GRAZING EMISSION SPECTROSCOPY

R.S. Becker, J.A. Golovchenko, and J.R. Patel first introduced the grazing emission geometry in 1983.[182] In GEXRF, the XES intensity is no longer monitored as a function of the incidence angle of the beam used for excitation but it is recorded as a function of the grazing emission angle (Figure 3.15). The grazing emission angle is varied around the critical angle, which depends in GEXRF, in addition to the sample matrix, on the energy of the emitted X-rays. Indeed, the radiation used to excite the XES signal is usually not incident at grazing conditions. A GEXRF setup is therefore sometimes called an *inverted GIXRF* setup since the paths for the excitation and the detection of the XES intensity are exchanged. In GIXRF, the refraction of the incident, and in GEXRF, the refraction of the emitted X-rays at the sample-environment interface and eventual intrasample interfaces have to be considered.

The physical equivalency between GIXRF and GEXRF can be explained by the principles of microscopic reversibility and reciprocity: if grazing incidence and grazing emission experiments were performed at the same X-ray energy, the distributions of the atoms contributing to the observed XES yield would be identical.[182] However, the X-ray energies of interest cannot be identical. Since the energy of the X-rays used to excite the XES signal is always larger than that of the emitted X-rays (respectively, in terms of wavelengths the incident wavelength is smaller than the emitted one), the critical angle will be smaller in GIXRF experiments than in GEXRF experiments. The angular scale is thus changed; if the scale was normalized to the critical angle, the XES yield on the respective grazing angle would look identical. However, the depth distribution of the atoms contributing to the recorded XES intensity will also differ due to different absorption coefficients, once the energy of interest is above the elemental absorption edge (GIXRF) and once below it (GEXRF). The reciprocity

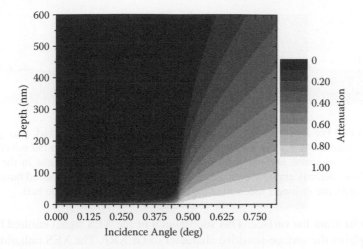

**FIGURE 3.18** Plot of the contribution to the detectable XES signal for the Fe Kα line emitted from a Fe bulk sample as a function of the emission angle and the depth (the distance from the sample surface). In contrast to GIXRF, the energy of interest is fixed, in the given example, the one of Fe Kα fluorescence line, and not variable (Figure 3.16). For angles below the critical angle for the considered energy ($\varphi_c = 0.474°$) solely the first 3–4 nm contribute significantly to the detectable XES yield, above the critical angle the depth range from which the XES signal can be recorded increases with the emission angle. The extinction depth, corresponding to the depth from which the emitted fluorescence intensity is attenuated by a factor of $e^{-1}$ on the way from the emitting atom to the surface, gives a good estimate of the depth range contributing to the measurement.

in optics, including the cases of stratified media and multilayer stacks, has been reviewed in the literature.[183]

However, in GEXRF, no external standing wave pattern is created since the incident X-rays impinge on the interface of the sample with the surrounding environment at angles far above the critical angle for total reflection.* Thus, only a fraction of a percent of the incident X-rays are reflected at the surface and the XES signal is also excited in the bulk volume of the sample. Nevertheless, for sufficiently small emission angles relative to the refracting surface, only the XES signal emitted by surface-near atoms is observable (see Figure 3.18). Indeed, X-rays emitted by atoms located far from the interface are considered as plane waves, which are refracted and reflected according to the Fresnel laws. Since the X-rays emitted by bulk atoms propagate at the sample interface into an optically denser medium (the surrounding environment), they are refracted away from the interface (inverted path with respect to the incident X-rays in GIXRF) and can consequently not be detected for grazing emission angles below the critical angle. The angular range up to the critical angle is indeed simply inaccessible for X-ray radiation (fluorescence and scattered X-rays)

---

* An internal standing wave pattern could nevertheless be created by a part of the XES signal if it was induced in a layer contained vertically by optically less dense layers (X-ray waveguide-like structures).

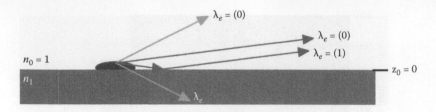

**FIGURE 3.19** For an emission angle smaller than the critical angle (dark gray paths in the figure), there are two possible detection paths due to the reflection on the substrate surface. Once the emission angle is above the critical angle (light gray paths in the figure), only the X-rays directly emitted toward the detector contribute to the measured fluorescence since the X-rays are no longer reflected at the surface and penetrate into the bulk.

produced far from the surface. This suppression of the XES signal emitted by bulk atoms explains the surface-sensitive character of GEXRF. The XES radiation emitted by near-surface atoms is described at the refracting interface by spherical waves, a Fourier decomposition of the latter shows that the corresponding evanescent wave has a nonnegligible amplitude.[118] Thus, GEXRF is characterized by the excitation of an internal evanescent wave, comparable to the external evanescent wave in GIXRF.

For grazing emission angles above the critical angle, the XES signal produced further away from the sample surface is also observable. Nevertheless, the probed depth region is still on the sub-micrometer scale since the shallow observation angles result in long emission paths within the sample and consequently in a quite pronounced absorption.

X-rays emitted at grazing angles with respect to a refracting interface can be either refracted upon their transmission through a refracting surface or they can be reflected. If, for example, the recorded XES signal originates from residues, grains, or particles on top of a reflecting surface (see Figure 3.19), the emitted X-rays can either reach the detector directly or via a reflection on the optically smooth and flat substrate surface. Thus, for emission angles below the critical angle, that is, in the angular regime where the surface reflectivity is close to 1, the probability to detect the XES signal is doubled. This probability enhancement for the detection of the XES signal is identical to the one in TXRF and GIXRF setups for the same type of sample.

A similar argument holds for thin layers deposited on a substrate with a larger refractive index for the XES signal of interest. It has already been explained that in GIXRF a standing wave pattern can be created inside a thin layer. In the case of GEXRF, if an atom at a height $z$ above the layer-substrate interface is considered, the difference in the path length for the emitted X-rays which were not reflected and for the ones which were reflected on the interface formed by the layer and the underlying substrate is $2 * z * \sin \varphi_e$ (Figure 3.20). Thus, depending on the emission angle, the X-rays emitted by atoms at the considered height, that is, contained in the same plane parallel to the refracting interfaces, can interfere constructively or destructively, depending if the path difference equals an even multiple or an odd multiple of the wavelength of the emitted X-rays. For a given emission angle, the conditions for constructive or destructive interference can be fulfilled at different,

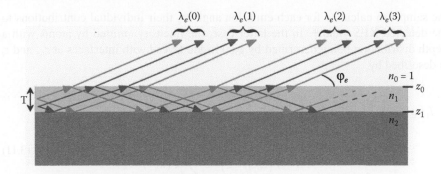

**FIGURE 3.20** Reflections on the two different interfaces of a layered sample result, for a given exit angle, in different paths for detection of X-rays of wavelength $\lambda_e$. The number in parentheses indicates the number of reflections, which X-rays following this path have undergone. Only for X-rays that are directly transmitted and reflected an even number of times have the length difference in the detection path independent on the depth position of the X-ray source.

periodic planes. The periodicity $D$ of the planes for which the coherently emitted X-rays interfere constructively (or destructively) depends on the emission angle and the fluorescence wavelength

$$D = \frac{\lambda_e}{2 * \sin\left(\varphi_e\right)}. \tag{3.10}$$

Thus, the periodicity presents a similar dependence as the periodicity of the nodal and antinodal lines in the grazing incidence geometry. The interference effects resulting from emission paths including an odd number of reflections can, however, not be observed in the dependence of the XES intensity on the grazing emission angle. Indeed, the interferences are washed out by X-rays emitted by atoms located in-between the periodic planes where the conditions for constructive or destructive interference are fulfilled. Nonetheless, interference fringes have been observed in measured angular XES intensity profiles, for example, for the Cr-K$\alpha$ line of Cr/Au/Cr layers deposited on quartz glass[184] and for the Zn K$\alpha$ line emitted from Zn films on a Au substrate,[185] and interpreted as an experimental evidence of the reciprocity theorem. The observed interference fringes can be explained by the following argument: the difference in the path length for emission paths including no reflection (i.e., directly transmitted X- rays) or an even number of reflections on the refracting interfaces is independent of the vertical position of the emitting atoms (Figure 3.20); thus, for X-rays emitted along one of these detection paths, the condition for constructive or destructive interference is independent of the position of the emitting atom.

The dependence of the XES intensity on the grazing emission angle can either be calculated with a matrix formalism to profit from the mathematical properties of the Hessenberg matrix for simulating the propagation of an electromagnetic wave through a stratified medium[186] or by considering directly the radiating sources emitting the XES signal of interest (wavelength $\lambda_e$) and their distribution in the sample to calculate for each emission angle $\varphi_e$ their individual contributions to the detected XES yield.[187] In the latter case, the intensity emitted by atoms with a depth distribution being described by $g_l(z)$ inside layer $l$ with interfaces at $z_{l-1}$ and $z_l$ is described by

$$
I_l(\varphi_e) = \prod_{n=0}^{l-1}\left(\frac{k_{n+1,z}}{k_{n,z}}*t_m^u\right)^2
$$

$$
\times \int_{z_l}^{z_{l-1}} g_l(z) \times e^{-2*\Im(k_{l,z})*(z_{(l-1)}-z)-2*\sum_{n=1}^{l-1}\Im(k_{n,z})*T_n} \times |\chi_e + \chi_o|^2 dz,
$$

(3.11)

where

$$
k_{l,z} = \frac{2*\pi}{\lambda_e}\sqrt{n_l^2 - \cos^2(\varphi_e)}.
$$

(3.12)

The factor preceding the integral takes into account the change in the strength of the electric wave field upon the transition of the different interfaces above the considered layer (if there is none this factor will be equal to 1), the exponential takes into account the absorption of the emitted XES signal in layer $l$ and the layers above it (the layer thickness being represented by $T_n$), and the last factor in the integral stands for interferences due to multiple reflections. The $z$-axis points out of the sample and its origin is on the interface formed by the sample with the surrounding environment. Eventually, different probabilities for the excitation of the XES signal have to be considered in $g_l(z)$ if, for example, the absorption in the depth direction of the probing beam changes significantly within the considered layer.

The transmission and reflection coefficients (Figure 3.21) are defined as

$$
t_l^u = \frac{2*k_{l,z}}{k_{l+1,z}+k_{l,z}} = 1-r_l^d = 1+r_l^u = \frac{k_{l,z}}{k_{l+1,z}}*t_l^d .
$$

(3.13)

For interfaces with a nonnegligible roughness, the transmission and reflection factors are attenuated by an exponential factor which depends on the coherence length.[24]

As illustrative examples, the calculated dependence of the XES intensity on the grazing emission angle of the Cr K$\alpha$ (E = 5414.7 eV) line for Cr particles having a spherical shape but different diameters (Figure 3.22b), respectively, for Cr layers of different thicknesses on the top of a Si layer (Figure 3.23b), for a Cr bulk sample (Figure 3.24), and for the P K$\alpha$ (E = 2013.7 eV) line of P ions implanted with different energies into a Si matrix (Figure 3.25a) are shown. In all the displayed examples, the XES intensities have been normalized to 1 at an emission angle of 1.5° in order to emphasize the dependence of the XES intensity on the emission angle.

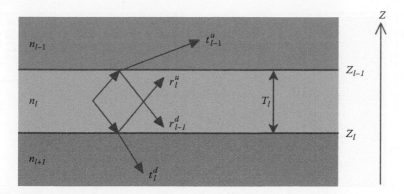

**FIGURE 3.21**  The definition of the reflection and transmission coefficients depends on the incidence direction of the X-rays on the interface between layer $l$ and the neighboring layers. In the considered example the optical density of the layers increases from top to bottom. The lower index indicates which interface is considered and the upper index indicates if the X-ray photon incident on the interfaces is propagating upward or downward.

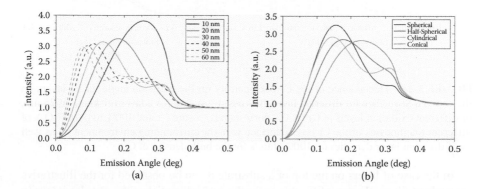

**FIGURE 3.22**  Dependence of the Cr K$\alpha$ intensity on the emission angle for Cr particles on the top of a Si surface for spherical particles of different thicknesses (a) and for 25 nm thick particles of different shapes (b). The highest XES yields are, because of the reflection of the Cr K$\alpha$ X-ray photons on the underlying substrate (Figure 3.19), observed for emission angles below the critical angle of Si for the energy of the Cr K$\alpha$ X-ray photons ($\varphi_c = 0.333°$). The angular intensity profiles depend on the size of the particles and their shape but also on the underlying substrate. The special case of particles located near the substrate edge is discussed in Bekshaev, A. et al., 2001, *Spectrochim. Acta B*, 56, 2385–2395.[188]

For particles on the top of a reflecting substrate, as illustrated for Cr particles on the top of Si (Figure 3.22a), the particle size is reflected in the evolution of the angular intensity profile. Likewise, the shape of the particles modify the dependence of the XES intensity on the emission angle (Figure 3.22b). This allows for distinguishing between particles of different volumes and surface areas. The special case of particles located near the substrate edge is discussed in Bekshaev, de Hoog, and van Grieken.[188]

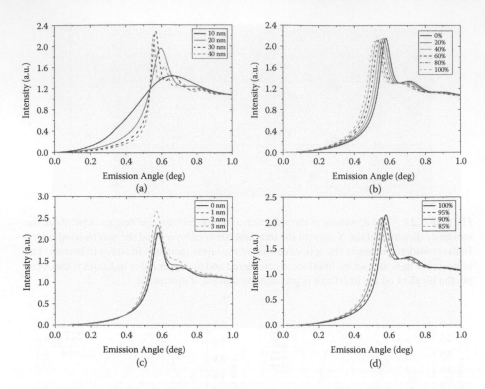

**FIGURE 3.23** Dependence of the Cr Kα intensity on the emission angle for Cr layers on the top of a Si surface for different layer thicknesses (a) and, for a layer thickness of 25 nm, of different oxidation levels of Cr (b: 0% corresponds to pure Cr and 100% to pure $CrO_2$), of different roughnesses of the Cr layer interface with the surrounding environment (c) as well as of different layer densities (d: 100% stands for the bulk density of Cr).

In the case of layers on the top of a substrate it can be observed for the illustrative example of Cr on the top of Si, that as in the case of particles, the angular intensity profile depends pronouncedly on the layer thickness (Figure 3.23a). In contrast to the preceding example, the highest XES yields are observed for emission angles above the critical angle of Cr for the Cr Kα line ($\varphi_c = 0.531°$). From the position of the critical angle, the layer density can be deduced (Figure 3.23d), while the observed fringes in the intensity profile allow for deducing the thickness.

The intensity at the largest emission angles depends merely on the amount of Cr atoms contained within the layer. Moreover, for layered samples the chemical state (in the given example, the oxidation of the Cr layer, Figure 3.23b) and the roughness (Figure 3.23c) also influence the angular intensity profile. The oxidation of Cr can be taken into account in the decrement factor and the imaginary part of the complex refractive index. This has been demonstrated for Al on top of Si in Kayser et al.[189] The distinction between Cr and $CrO_2$ is not possible in the case of Cr particles on the top of a substrate since the only refracting interface is the one formed by the substrate and the surrounding environment. Thus, the change in the refractive index

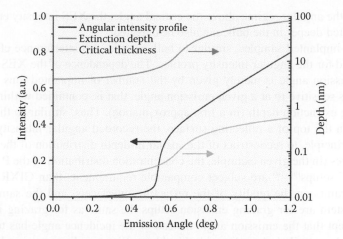

**FIGURE 3.24** Dependence of the Cr Kα intensity on the emission angle for a Cr bulk sample. In contrast to the layered samples, the XES intensity increases in the angular regime above the critical angle since an increased number of Cr atoms can contribute to the detected XES intensity. The extinction depth and the critical thickness are displayed as well for the given example.

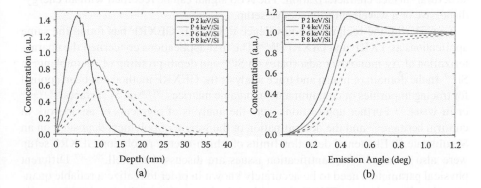

**FIGURE 3.25** Calculated depth concentration distribution of P ions implanted at different energies in a Si matrix (a) and the dependence of the P Kα line on the emission angle for the corresponding depth distributions. Reversely it is possible to retrieve the concentration distribution of the P ions from the angular intensity profile.

caused by the oxidation on Cr is not visible and the oxidation of Cr has to be assessed by other means, for example, the energy of the XES line or XAS measurements.

For a bulk Cr sample, the angular intensity dependence of the Cr Kα line on the emission angle differs from the particle-like and the layer-like distributions (Figure 3.24). Indeed, the shape of the angular intensity profile is mainly influenced by the absorption of the emitted photons and also by the refraction at the interface formed by the sample and the surrounding environment but not by interference effects. In the angular regime above the critical angle, the XES intensity emitted by atoms located deeper in the bulk can also be detected.

For ion-implanted samples, similar to bulk samples, no interference effects can be observed for the angular intensity profiles. The dependence of the XES intensity on the emission angle is merely given by the number of implanted ions to which the setup is sensitive to at a given emission angle, that is, contained within the corresponding extinction depth (in a first approximation). Thus, similar to the case of particles on the top of a reflecting surface, the recorded angular intensity profiles allow in principle, to reconstruct of the spatial in-depth distribution of the emitting XES sources (in the given example, the concentration distribution of the P ions).

GEXRF setups[190–193] are subject comparable requirements than GIXRF setups. The constraints on the quality of the refracting interface(s) and the sample positioning system are for grazing emission setups the same as for grazing incidence setups except that the emission angle instead of the incidence angle has to be well defined and controlled. An adjustable (double) collimator slit system and/or a dispersive device, for example, a grating or a crystal, can be used to define the grazing emission angle with respect to the sample surface. The angular resolution depends on the angular acceptance of the detection setup and its separation distance to the irradiated sample area. Both parameters can be adapted to the experimental needs and purposes: detection sensitivity for trace-element control or angular accuracy for structural surface characterization. The XES signal can be recorded with an energy-dispersive or a wavelength-dispersive setup.

Considering the reciprocal equivalence to GIXRF, GEXRF has found the same applications as TXRF and GIXRF.[118,194] The first applications concerned the characterization of Ag monolayer adsorbates on Si[195] and depth-profiling of As implants in Si.[196] In the domain of micro and trace analysis the GEXRF method has been applied for tracing impurities or contaminants in organic matrices,[197,198] on Si surfaces,[118,199,200] or in water.[201] Further applications were the analysis of pigments of an artwork in cultural heritage[202] and the determination of the weight of polymers deposited on an Si substrate.[203] Elemental detection limits of a high-energy resolution GEXRF setup were also assessed.[204] Quantification issues are discussed as well.[205–207] Different physical parameters need to be accurately known in order to realize a reliable quantification. In analogy to TXRF, the critical thickness, that is, the maximum sample thickness for which absorption effects can be safely neglected, was also introduced in GEXRF.[206] Special substrates with a periodic structure were proposed to enhance the sensitivity.[208] If only elemental or chemical information is required, a measurement at a single emission angle is sufficient, in the same way as in TXRF the isoradiant angle can be chosen to measure the XES intensity.[206]

Recording the dependence of the XES intensity allows on the other hand to obtain besides the elemental information also structural information on thin films, for example, in order to study surface oxidation or the density and thickness of the deposited films.[209–211] A depth-dependent probing by varying the grazing emission angle has been proposed to perform chemical speciation with soft X-rays for a three-layered sample.[212] Al layers deposited on Si were investigated regarding their thickness, oxidation level, roughness, and density for layers with different nominal thicknesses ranging from very thin to medium values.[189] The thickness, density, and composition of different nitride films were also characterized.[213,214] The observed angular intensity profile has been shown to not only depend on the layer from which the XES signal is studied but also on the neighboring layers,[215] allowing reversely to characterize the latter in principle indirectly. On a bulk Fe-Cr-Ni alloy sample, the surface oxidation and the width of the Cr depletion zone formed upon the oxidation were studied.[216] As in GIXRF, the depth profiling of ion-implanted wafers is also reported.[217]

### 3.5.4 EXPERIMENTAL DIFFERENCES BETWEEN GIXRF AND GEXRF

Although GIXRF and GEXRF are based on the same physical principles and through the reciprocity theorem are also physically equivalent, some experimental differences have to be considered.

It has already been pointed out that the critical angle referred to in GEXRF is larger than in GIXRF since it is defined relative to the wavelength of the XES signal and not to that of the incident X-rays used to excite the XES signal. This mainly results in a rescaled angular range with respect to the critical angle. Also, due to the different X-ray energies of interest, once below (GEXRF), once above (GIXRF) the absorption edge, the sensitivity toward the sample matrix is different.

A further experimental difference is given by the geometrical arrangement of GIXRF and GEXRF. In GIXRF, for geometrical reasons, the surface region over which the standing wave pattern extends depends inversely on the incidence angle. Thus, the incident beam spreads over a wide sample surface. The irradiated area depends strongly on the incidence angle. Also, laterally resolved studies with a resolution in the scale of a few micrometers are difficult to realize in GIXRF, in contrast to GEXRF. Indeed, in GEXRF the incidence angle of the X-ray photons used for the excitation of the XES signal is chosen, if possible, to be perpendicular to the sample surface. Thus, if microsized X-ray beams (collimated or focused X-ray beams) are used, surface mapping applications are quite straightforward to realize.[190,204,218,219] Consequently, GEXRF presents the potential for three-dimensional analysis,[220,221] the in-depth resolution being on the nanometer scale and the lateral resolution being defined by the irradiated sample area.

A further experimental aspect that has to be considered is the type of ionizing radiation which can be used to excite the XES signal. While by definition in grazing incidence techniques a (monochromatic) X-ray beam is required to excite the XES signal, GEXRF setups are less restrictive toward the exact production process of the XES signal since the experimental results depend mainly on the refraction of the emitted photons. In the previously reported applications X-ray tubes or a synchrotron radiation source were employed to excite the XES signal. However, other excitation sources were explored as well: electron beams,[222–224] proton beams,[225,226] and a laser plasma source.[227] The advantage of using charged particles is that the fluorescence signal is mainly produced close to the surface, that is, in the sample volume to which the GEXRF measurement is sensitive, due to the lower penetration depth of particles. Disadvantages are the inherent Bremsstrahlung background and, depending on the sample to be analyzed, a depth-dependent excitation yield of the XES signal.

A further major experimental difference between grazing incidence and grazing emission concerns the excitation and detection efficiency of the XES signal. Indeed, the grazing incidence angles entail an excitation of the XES signal from an increased number of surface-near atoms because of the large beam footprint. Furthermore, energy-dispersive detectors mounted close to the sample surface subject to the measurement ensure a large solid angle of detection. In the grazing emission geometry, the unavoidable collimation of the emitted XES signal needed for the definition of the grazing emission angle reduces the setup luminosity. As a consequence, grazing incidence setups offer better direct detection limits than grazing emission setups, the difference being about one order of magnitude,[130,228] although on a theoretical basis, considering the detection efficiency and signal-to-noise ratios, similar detection limits are expected.[229] If preconcentration techniques can be employed, TXRF and GEXRF should present similar detection limits.[199,204] In the domain of depth profiling, a comparison between GIXRF and GEXRF for Al implantations in Si can be found in Hönicke et al.[230]

However, the necessary collimation of the XES signal in grazing emission setups also allows combining them with wavelength-dispersive setups without being further penalized by their small solid angle of detection. Indeed, in the latter a collimation of the XES radiation subject to the measurement is inherently realized by the dispersive element, which separates the emitted X-ray photons by their wavelength. Thus, a combination of GEXRF and a wavelength-dispersive detection setup is quite straightforward. It can then be stated that GEXRF corresponds to a special configuration of a WDXRF setup, similar to the case of TXRF/GIXRF and EDXRF setups.

The main advantages offered by wavelength-dispersive setups are their energy resolution, providing some sensitivity to chemical states and a good separation of the many L-lines of mid-Z elements and/or M-lines of heavy elements, their good background rejection capabilities, which contributes to an improved signal-to-background ratio, and finally their sensitivity toward low-Z elements. The low-Z sensitivity was one of the main motivations to develop wavelength-dispersive GEXRF spectrometers as a complement to the grazing incidence setups used for metallic contamination control.[194] The background rejection capabilities are especially advantageous if the benefits of synchrotron radiation are profited from at the same time.[231] Moreover, the potential of XES for element-specific chemical speciation can only be fully exploited with synchrotron radiation and a high-energy resolution detection setup. GEXRF measurements with a high-energy resolution von Hamos crystal X-ray spectrometer in combination with synchrotron radiation have been reported.[189,199,204,217] Wavelength-dispersive detection setups were also combined with the grazing incidence geometry[166,232] at the expense of a decreased detection efficiency since in addition to the incidence, the emitted radiation was also collimated.

However, wavelength-dispersive setups are not suitable for the elemental determination of an unknown sample since the alignment of the dispersion device and the detector and consequently the covered energy range, which is much narrower than with energy-dispersive detection setups, depend on the wavelength of the emitted X-rays being measured. In this perspective, the combination of energy-dispersive detectors with the grazing emission geometry can also be realized.[130] The solid angle for detection will not be significantly increased, but the intensity dependence on the grazing emission angle of different emission lines of one or more elements can be recorded simultaneously (and not separately as with a wavelength-dispersive detector). Thus, it is possible to acquire for one sample, different data sets without changes in the experimental setup.[118]

Finally, the combination of the grazing incidence and the grazing emission geometry has also been realized. Either the dependence of the XES intensity on the incidence angle was measured for different emission angles or vice versa. Due to the collimation of the incident and emitted X-ray radiation, the detection efficiency of these setups is reduced. The main motivation for these setup types was the very low background since the excitation and the detection process focus both on the near-surface region. Furthermore, varying the incidence and the emission angle, and consequently the observation depth, allows acquiring multiple data sets for the same sample. Mostly the density and thickness of thin layers were characterized with setups where the grazing incidence and grazing emission geometries were realized simultaneously. Moreover diffusion processes, chemical conditions, and the surface/interface roughness were studied.[185,233,234]

In Sasaki et al.,[185] the intensity dependence of the XES signal on the grazing emission angle was assessed by means of an imaging plate placed at a sufficient distance from the samples in a single measurement and not in a sequence of measurements at different sample, respectively, detector positions. Thus, a spatially resolved measurement allows acquiring the XES intensity dependence on the grazing emission angle in a single measurement without moving any components in order to change the sample-detector arrangement. This has also been demonstrated with a two-dimensional position-sensitive area detector,[235] the grazing emission angles being dispersed along one of the detector dimensions while the second detector dimension permits increasing the solid angle of detection for each grazing emission angle. A scanning-free arrangement is especially advantageous with pulsed, broadband sources like X-ray free electron lasers (XFELs) or in combination with other scanning-based X-ray techniques like XAS. The feasibility of XAS measurements with a scanning-free GEXRF arrangement was demonstrated in Shinoda et al.[236,237] The fact that the different emission angles do not need to be scanned sequentially presents a considerable gain in time, compensating for the smaller solid angle of detection of GEXRF setups. Also, the fact that different sample depths are probed at once is certainly an advantage for time-resolved measurements. In addition the experiment should be less sensitive to thermal drifts, vibrational instabilities over time, and intensity fluctuations of the XES excitation source. A much simpler sample positioning system can be used, only a reference holder being required. In contrast to GEXRF, there is no alternative to point-by-point measurements at different angular positions in GIXRF since the XES signals excited simultaneously by X-rays with different incidence angles on the same sample spot cannot be separated from each other.

### 3.5.5 COMBINATIONS WITH OTHER TECHNIQUES

Both GIXRF and GEXRF are XES techniques, which in addition to the XES-specific applications offer the possibility for surface-sensitive and depth-resolved studies. However, in contrast to XAS, for chemical speciation by means of RXES, a wavelength-dispersive setup is required. In this case, a combination with the grazing emission geometry is more favorable. GIXRF and GEXRF can also be used for chemical speciation if the difference in the refractive index of the considered chemical compounds for the energy of interest is sufficiently important to detect a change in the critical angle.[189] Nevertheless, the surface-sensitive character and depth-profiling possibilities offered by the grazing incidence and grazing emission geometries can also be exploited in combination with other X-ray techniques in order to gather complementary or additional information such as the unoccupied density of electronic states or the local atomic/molecular structure and configurations. For example, diffraction and scattering techniques were used in the grazing incidence or grazing emission geometry. A setup combining different grazing incidence techniques, allowing for complementary characterization by combining the information gathered with each separate technique and thus profiting from the advantages offered by the individual methods, is described in Holfelder et al.[238]

A straightforward combination of different grazing incidence techniques is that of GIXRF and X-ray reflectivity (XRR) or X-ray diffuse scattering in grazing incidence.[147] In GIXRF and XRR, a dependence on the incidence angle of the incident monochromatic X-rays is assessed; in GIXRF, the intensity of the XES signal, and in XRR, the intensity of the specular reflection from the sample surface. Both techniques can be implemented in a single setup. XRR by itself does also allow for the characterization of thin layers (as noted in a number of examples[239–241]) if the studied layer is sufficiently thick and laterally homogeneous. GIXRF in combination with XRR was used to characterize $CoPt_3$ nanoparticle films on the top of an Au substrate,[242] Fe diffusion in amorphous Si,[243] and the interfaces of a Ti/Ni/Ti trilayer.[244] Contrary to GIXRF, in XRR the dependence of the specular reflected X-ray intensity on the incidence angle can be recorded at once by deflecting the probing beam with a dispersive element onto the sample surface in order to simultaneously cover a range of incidence angles and by a spatially resolved detection of the reflected X-rays.[245] Grazing incidence beams are, moreover, well suited for XPS since the penetration of the incident beam and, thus, the depth region in which atoms are ionized matches the escape depth of the photoelectrons.[246]

X-ray absorption measurements were realized in both grazing geometries by recording the XES intensity (or equivalently the total fluorescence yield) as a function of the energy of the incident X-ray beam. By means of GIXRF, such measurements were performed on model films for thin film solar cells[247,248] and on two-layered (boron-carbon-nitride film on the top of Ni) thin film systems.[249] Two different geometries to acquire the fluorescence yield in an absorption measurement in grazing incidence geometry are compared in Maurizio et al.[250] while grazing incidence absorption (and emission) spectroscopy in the domain of geochemistry is discussed in Waychunas.[251] However, for layered samples, in absorption measurements performed in grazing incidence geometry the incidence angle needs to be varied as a function of the incidence energy in order to keep a constant probing depth[252] (see Figure 3.16 for a Fe sample and energies around the Fe K-edge). Indeed, the critical angle in addition to the elemental composition of a layer depends on the energy of the incident radiation. Around an elemental absorption edge the critical angle and, thus, the probed depth region vary quite significantly because of a dip in the energy dependence of the decrement factor for the considered layer. For particles on the top of a reflecting surface, this effect is less important since only the standing wave field will be slightly modulated when scanning the incidence energy.

In contrast to GIXRF, absorption measurements in grazing emission geometry can, independently of the sample, be performed at a constant grazing emission angle since in GEXRF the critical angle and, thus, the probed depth region depend solely on the energy of the measured XES signal. Since the probed depth region below the critical angle is only a few nanometers deep, absorption measurements can be realized on the surface-near atoms and the result reflects directly the dependence of the photoelectric cross section on the beam energy.[253] Arsenic traces on top of Si were investigated with absorption measurements in grazing emission geometry and compared to measurements in grazing incidence geometry.[218] Moreover, depth-resolved absorption measurements were realized by means of GEXRF on Fe-Mn alloys,[236] electrode/electrolyte interfaces,[254] and oxygen electrodes of solid-oxide fuel cells[237]; a scanning-free arrangement allowed to record the XES intensity simultaneously for different grazing emission angles, the monitored angular range being above the critical angle.[236,237]

### 3.5.6   CONCLUSIVE REMARKS

While RXES is a very sensitive technique for probing the chemical configuration of an element by studying the X-rays emitted during an electronic transition affecting a valence shell, the grazing incidence and grazing emission geometries provide a tunable surface sensitivity ranging from a few nanometers to several hundred nanometers. In combination with a grazing geometry, RXES experiments can be realized *in situ* on the surface, that is, the region in which chemical reactions take place. Moreover, the probe of interest can be studied in a nonpreparative and noninvasive way. In particular, the emission geometry in combination with a wavelength-dispersive detection setup allows for a reliable chemical characterization of the surface-near atoms by measuring the XES signal at an adequate grazing emission angle, the high-energy resolution being mandatory for the detection of energy shifts in the XES signal caused by the evolution from one chemical configuration to another. Moreover, the XES studies do not require any scanning components as, for example, XAS studies require. Studying the dependence of the XES intensity on the incidence, respectively, the emission angle also allows characterizing the spatial distribution of the studied species, respectively, to focus on different intrasample interfaces or volumes. In the grazing emission geometry this can even be realized without scanning through motor positions in order to change the sample-detector variation by using a position-sensitive detector.[134] This could be of special importance for single-shot experiments.

A step further would be marked by time-resolved studies which would allow following the temporal evolution of the studied system.[255,256] This would be of special importance in the study of dynamical systems under operating conditions.

## ACRONYM LIST

**AES:** Auger Electron Spectroscopy
**EDXRF:** Energy-Dispersive X-ray Fluorescence
**GEXRF:** Grazing Emission X-ray Fluorescence
**GIXRF:** Grazing Incidence X-ray Fluorescence
**HEROS:** High-Energy Resolution Off-Resonant Spectroscopy
**HERFD-XAS:** High-Energy Resolution Fluorescence Detected XAS
**MEIS:** Medium Energy Ion Scattering
**RIXS:** Resonant Inelastic X-ray Spectroscopy
**RXES:** Resonant X-ray Emission Spectroscopy
**SEM:** Scanning Electron Microscopy
**SIMS:** Secondary Ion Mass Spectroscopy
**STM:** Scanning Tunneling Microscopy
**TEM:** Transmission Electron Microscopy
**TOF-SIMS:** Time-of-Flight-SIMS
**TXRF:** Total Reflection X-ray Fluorescence
**WDXRF:** Wavelength-Dispersive X-ray Fluorescence
**XAS:** X-ray Absorption Spectroscopy
**XES:** X-ray Emission Spectroscopy
**XFEL:** X-ray Free Electron Laser
**XPS:** X-ray Photoelectron Spectroscopy
**XRF:** X-ray Fluorescence
**XRR:** X-ray Reflectivity
**XSW:** X-ray Standing Wave

## REFERENCES

1. Yeh, J.J. and I. Lindau. Atomic subshell photoionization cross sections and asymmetry parameters: $1 \leqslant Z \leqslant 103$. *Atomic Data and Nuclear Data Tables* 32, (1985) 1–155.
2. Hubbell, J.H. and I. Øverbø. Relativistic atomic form factors and photon coherent scattering cross sections. *J. Phys. Chem. Ref. Data* 8, (1979) 69.
3. Hubbell, J.H. et al. Atomic form factors, incoherent scattering functions, and photon scattering cross sections. *J. Phys. Chem. Ref. Data* 4, 3, 4, (1975) 471–538.
4. Kotani, A. and S. Shin. Resonant inelastic X-ray scattering spectra for electrons in solids. *Rev. Mod. Phys.* 73, (2001) 203–246.
5. Bergmann, U. and P. Glatzel. X-ray emission spectroscopy. *Photosynth. Res.* 102, (2009) 255–266.
6. Kotani, A. Core-hole effect in the Ce L-3 X-ray absorption spectra of CeO2 and CeFe2: New examination by using resonant X-ray spectroscopy. *Modern Physics Letters B* (2013) 27.
7. Glatzel, P., U. Bergmann, F.M.F. de Groot, B.M. Weckhuysen, and S.P. Cramer. A study of transition metal K absorption pre-edges by resonant inelastic X-ray scattering (RIXS). *Phys. Scripta* T115, (2005) 1032–1034.
8. Glatzel, P. et al. Hard X-ray photon-in-photon-out spectroscopy with lifetime resolution—Of XAS, XES, RIXSS and HERFD. *SRI* (2007)1–4.
9. Vankó, G. et al. Spin-state studies with XES and RIXS: From static to ultrafast. *Journal of Electron Spectroscopy and Related Phenomena* 188, (2013) 166–171.

10. de Groot, F.M.F. Site-selective XAFS: A new tool for catalysis research. *Top Catal.* 10, (2000) 179–186.

11. Marchenko, T. et al. Resonant inelastic X-ray scattering at the limit of subfemtosecond natural lifetime. *J. Chem. Phys.* 134, (2011) 144308.

12. Singh, J., C. Lamberti, and J.A. Van Bokhoven. Advanced X-ray absorption and emission spectroscopy: *In situ* catalytic studies. *Chemical Society Reviews* 39, (2010) 4754–4766.

13. Glatzel, P. et al. Reflections on hard X-ray photon-in/photon-out spectroscopy for electronic structure studies. *Journal of Electron Spectroscopy and Related Phenomena* 188, (2013) 17–25.

14. Westre, T.E. et al. A multiplet analysis of Fe K-edge 1s->3d pre-edge features of iron complexes. *J. Am. Chem. Soc.* 119, (1997) 6297–6314.

15. Vedrinskii, R.V., V.L. Kraizman, A.A. Novakovich, P.V. Demekhin, and S.V. Urazhdin. Pre-edge fine structure of the 3d atom K X-ray absorption spectra and quantitative atomic structure determinations for ferroelectric perovskite structure crystals. *J. Phys. Condens. Mat.* 10, (1998) 9561–9580.

16. Yamamoto, T. Assignment of pre-edge peaks in K-edge X-ray absorption spectra of 3d transition metal compounds: Electric dipole or quadrupole? *X-ray Spectrom.* 37, (2008) 572–584.

17. Schwitalla, J. and H. Ebert. Electron core-hole interaction in the X-ray absorption spectroscopy of 3d transition metals. *Phys. Rev. Lett.* 80, (1998) 4586–4589.

18. Cabaret, D., A. Bordage, A. Juhin, M. Arfaoui, and E. Gaudry. First-principles calculations of X-ray absorption spectra at the K-edge of 3d transition metals: An electronic structure analysis of the pre-edge. *Phys. Chem. Chem. Phys.* 12, (2010) 5619.

19. Glatzel, P., M. Sikora, and M. Fernández-García. Resonant X-ray spectroscopy to study K absorption pre-edges in 3d transition metal compounds. *Eur. Phys. J. Spec. Top.* 169, (2009) 207–214.

20. Glatzel, P. and U. Bergmann. High resolution 1s core hole X-ray spectroscopy in 3D transition metal complexes—Electronic and structural information. *Coordination Chemistry Reviews* 249, (2005) 65–95.

21. Wu, Z.Y., G. Ouvrard, P. Gressier, and C.R. Natoli. Ti and O K edges for titanium oxides by multiple scattering calculations: Comparison to XAS and EELS spectra. *Phys. Rev. B* 55, (1997) 10382–10391.

22. Grunes, L.A. Study of the K edges of 3d transition-metals in pure and oxide form by X-ray-absorption spectroscopy. *Phys. Rev. B* 27, (1983) 2111–2131.

23. Campbell, J.L. and T. Papp. Atomic level widths for X-ray spectrometry. *X-ray Spectrom.* 24, (1995) 307–319.

24. Campbell, J.L. and T. Papp. Widths of the atomic K-N7 levels. *Atomic Data and Nuclear Data Tables* 77, (2001) 1–56.

25. Keski-Rahkonen, O. and M. Krause. Total and partial atomic-level widths. *At. Data Nucl. Data Tables* 14, (1974) 139–146.

26. Sokaras, D. et al. A seven-crystal Johann-type hard X-ray spectrometer at the Stanford Synchrotron Radiation Lightsource. *Rev. Sci. Instrum.* 84, (2013) 053102.

27. Kleymenov, E. et al. Five-element Johann-type X-ray emission spectrometer with a single-photon-counting pixel detector. *Rev. Sci. Instrum.* 82, (2011) 065107.

28. Sokaras, D. et al. A high resolution and large solid angle X-ray Raman spectroscopy end-station at the Stanford Synchrotron Radiation Lightsource. *Rev. Sci. Instrum.* 83, (2012) 043112–043112–9.

29. Alonso-Mori, R. et al. A multi-crystal wavelength dispersive X-ray spectrometer. *Rev. Sci. Instrum.* 83, (2012) 073114–073114–9.

30. Verbeni, R., M. Kocsis, S. Huotari, and M. Krisch. Advances in crystal analyzers for inelastic X-ray scattering. *Journal of Physics and Chemistry of Solids* 66, (2005) 2299–2305.

31. Szlachetko, J. et al. A von Hamos X-ray spectrometer based on a segmented-type diffraction crystal for single-shot X-ray emission spectroscopy and time-resolved resonant inelastic X-ray scattering studies. *Rev. Sci. Instrum.* 83, (2012) 103105–103105–7.
32. Kavčič, M. et al. Design and performance of a versatile curved-crystal spectrometer for high-resolution spectroscopy in the tender X-ray range. *Rev. Sci. Instrum.* 83, (2012) 033113.
33. Sá , J. et al. Direct observation of charge separation on Au localized surface plasmons. *Energy Environ. Sci.* 6, (2013) 3584–3588.
34. Deslattes, R.D. et al. X-ray transition energies: New approach to a comprehensive evaluation. *Rev. Mod. Phys.* 75, (2003) 35–99.
35. Bohinc, R. et al. Dissociation of chloromethanes upon resonant sigma* excitation studied by X-ray scattering. *Journal of Chemical Physics* 139, (2013).
36. Morin, P. and I. Nenner. Atomic autoionization following very fast dissociation of core-excited HBr. *Phys. Rev. Lett.* 56, (1986) 1913–1916.
37. Aksela, H. et al. Decay channels of core-excited HCl. *Phys. Rev. A* 41, (1990) 6000–6005.
38. Morin, P. and C. Mirone. Ultrafast dissociation: An unexpected tool for probing molecular dynamics. *Journal of Electron Spectroscopy and Related Phenomena* (2012) 259–266.
39. Miron, C., P. Morin, D. Céolin, L. Journel, and M. Simon. Multipathway dissociation dynamics of core-excited methyl chloride probed by high resolution electron spectroscopy and Auger-electron–ion coincidences. *J. Chem. Phys.* 128, (2008) 154314.
40. Travnikova, O. et al. On routes to ultrafast dissociation of polyatomic molecules. *J. Phys. Chem. Lett.* 4, (2013) 2361–2366.
41. Simon, M. et al. Femtosecond nuclear motion of HCl probed by resonant X-ray Raman scattering in the Cl 1s region. *Phys. Rev. A* 73, (2006) 020706.
42. Gelmukhanov, F. and H. Ågren. X-ray resonant scattering involving dissociative states. *Phys. Rev. A* 54, (1996) 379–393.
43. Gelmukhanov, F., P. Sałek, T. Privalov, and H. Ågren. Duration of X-ray Raman scattering. *Phys. Rev. A* 59, (1999) 380–389.
44. Kavčič, M. et al. Electronic state interferences in resonant X-ray emission after k-shell excitation in HCl. *Phys. Rev. Lett.* 105, (2010) 113004.
45. Guillemin, R. et al. Angular and dynamical properties in resonant inelastic X-ray scattering: Case study of chlorine-containing molecules. *Phys. Rev. A* 86, (2012) 013407.
46. Nahle, A., F.C. Walsh, C. Brennan, and K.J. Roberts. X-ray design constraints for *in situ* electrochemical cells: Importance of window material, electrolyte, and X-ray wavelength. *J. Appl. Crystallogr.* 32, (1999) 369–372.
47. De Marco, R. and J.-P. Veder. *In situ* structural characterization of electrochemical systems using synchrotron-radiation techniques. *TrAC Trends in Analytical Chemistry* 29, (2010) 528.
48. Bergmann, U., C.R. Horne, T.J. Collins, J.M. Workman, and S.P. Cramer. Chemical dependence of interatomic X-ray transition energies and intensities—A study of Mn K β and K β(2,5) spectra. *Chemical Physics Letters* 302, (1999) 119–124.
49. Bergmann, U., J. Bendix, P. Glatzel, H.B. Gray, and S.P. Cramer. Anisotropic valence → core X-ray fluorescence from a [Rh(en)[sub 3]][Mn(N)(CN)[sub 5]]·H[sub 2]O single crystal: Experimental results and density functional calculations. *J. Chem. Phys.* 116, (2002) 2011.
50. Hayashi, H. et al. Selective XANES spectroscopy from RIXS contour maps. *J. Phys. Chem. Solids* 66, (2005) 2168–2172.
51. Safonova, O.V. et al. Identification of CO adsorption sites in supported Pt catalysts using high-energy-resolution fluorescence detection X-ray spectroscopy. *J. Phys. Chem. B* 110, (2006) 16162–16164.

52. Vankó, G. et al. Probing the 3d spin momentum with X-ray emission spectroscopy: The case of molecular-spin transitions. *J. Phys. Chem. B* 110, (2006) 11647–11653.

53. Lee, K.E., M.A. Gomez, T. Regier, Y. Hu, and G.P. Demopoulos. Further understanding of the electronic interactions between N719 sensitizer and anatase TiO 2films: A combined X-ray absorption and X-ray photoelectron spectroscopic study. *J. Phys. Chem. C* 115, (2011) 5692–5707.

54. Glatzel, P., J. Singh, K.O. Kvashnina, and J.A. Van Bokhoven. *In situ* characterization of the 5d density of states of Pt nanoparticles upon adsorption of CO. *J. Am. Chem. Soc.* 132, (2010) 2555–2557.

55. Sham, T.K., M.J. Ward, M.W. Murphy, L.J. Liu, and W.Q. Han. Pt L 3,2-edge whiteline anomaly and its implications for the chemical behaviour of Pt 5d 5/2 and 5d 3/2 electronic states—A study of Pt-Au nanowires and nanoparticles. *J. Phys.: Conf. Ser.* 430, (2013) 012018.

56. Szlachetko, J. and J. Sá. Rational design of oxynitride materials: From theory to experiment. *Cryst. Eng. Comm.* 15, (2013) 2583.

57. Kollbek, K. et al. X-ray absorption and emission spectroscopy of TiO2 thin films with modified anionic sublattice. *Radiation Physics and Chemistry* 93, (2013) 40–46. doi:10.1016/j.radphyschem.2013.03.035.

58. Sikora, M. et al. 1s2p resonant inelastic X-ray scattering-magnetic circular dichroism: A sensitive probe of 3d magnetic moments using hard X-ray photons. *J. Appl. Phys.* (2012) 111.

59. Sikora, M. et al. Strong K-edge magnetic circular dichroism observed in photon-in–photon-out spectroscopy. *Phys. Rev. Lett.* 105, (2010) 037202.

60. Kuehn, T.-J., W. Caliebe, N. Matoussevitch, H. Boennemann, and J. Hormes. Site-selective X-ray absorption spectroscopy of cobalt nanoparticles. *Applied Organometallic Chemistry* 25, (2011) 577–584.

61. Hayashi, H., T. Azumi, A. Sato, and Y. Udagawa. A cartography of Kβ resonant inelastic X-ray scattering for lifetime-broadening-suppressed spin-selected XANES of α-Fe2O3. *Journal of Electron Spectroscopy and Related Phenomena* 168, (2008) 34–39.

62. Kuehn, W., K. Reimann, M. Woerner, and T. Elsaesser. Phase-resolved two-dimensional spectroscopy based on collinear n-wave mixing in the ultrafast time domain. *J. Chem. Phys.* 130, (2009) 164503.

63. Szlachetko, J. et al. *In situ* hard X-ray quick RIXS to probe dynamic changes in the electronic structure of functional materials. *Journal of Electron Spectroscopy and Related Phenomena* 188, (2013) 161–165.

64. Hoszowska, J., J.C. Dousse, J. Kern, and C. Rhême. High-resolution von Hamos crystal X-ray spectrometer. *Nuclear Instruments and Methods in Physics Research Section A: Accelerators, Spectrometers, Detectors and Associated Equipment* 376, (1996) 129–138.

65. Miyamoto, T., H. Niimi, Y. Kitajima, T. Naito, and K. Asakura. Ag L 3-Edge X-ray absorption near-edge structure of 4d 10(Ag +) compounds: Origin of the edge peak and its chemical relevance. *J. Phys. Chem. A* 114, (2010) 4093–4098.

66. Hu, Z. et al. Multiplet effects in the Ru $L_{2,3}$ X-ray-absorption spectra of Ru(IV) and Ru(V) compounds. *Phys. Rev. B* 61, (2000) 5262–5266.

67. Pearson, D., C. Ahn, and B. Fultz. White lines and d-electron occupancies for the 3d and 4d transition metals. *Phys. Rev. B* 47, (1993) 8471–8478.

68. Jeon, Y., J. Chen, and M. Croft. X-ray-absorption studies of the d-orbital occupancies of selected 4d/5d transition metals compounded with group-III/IV ligands. *Phys. Rev. B* 50, (1994) 6555–6563.

69. Sham, T. L-edge X-ray-absorption systematics of the noble metals Rh, Pd, and Ag and the main-group metals In and Sn: A study of the unoccupied density of states in 4d elements. *Phys. Rev. B* 31, (1985) 1888–1902.

70. Nakazawa, M., K. Fukui, and A. Kotani. Theory of X-ray absorption and resonant X-ray emission spectra by electric quadrupole excitation in light rare-earth systems. *Journal of Solid State Chemistry* 171, (2003) 295–298.

71. De Crescenzi, M., M. Diociaiuti, P. Picozzi, and S. Santucci. Electronic and structural investigations of palladium clusters by X-ray absorption near-edge structure and extended X-ray absorption fine-structure spectroscopies. *Phys. Rev. B* 34, (1986) 4334–4337.

72. Liu, R., L.Y. Jang, H.H. Hung, and J. Tallon. Determination of Ru valence from X-ray absorption near-edge structure in RuSr2GdCu2O8-type superconductors. *Phys. Rev. B* 63, (2001) 212507.

73. Saes, M. et al. Observing photochemical transients by ultrafast X-ray absorption spectroscopy. *Phys. Rev. Lett.* 90, (2003) 047403.

74. Sá, J. et al. Magnetic manipulation of molecules on a non-magnetic catalytic surface. *Nanoscale* 5, (2013) 8462.

75. Sá, J. et al. Fine tuning of gold electronic structure by IRMOF post-synthetic modification. *RSC Adv.* 3, (2013) 12043.

76. Makosch, M. et al. HERFD XAS/ATR-FTIR batch reactor cell. *Phys. Chem. Chem. Phys.* 14, (2012) 2164.

77. Sá, J. et al. Evaluation of Pt and Re oxidation state in a pressurized reactor: Difference in reduction between gas and liquid phase. *Chemical Communications* 47, (2011) 6590.

78. Eisenberger, P., P. Platzman, and H. Winick. X-ray resonant raman scattering: Observation of characteristic radiation narrower than the lifetime width. *Phys. Rev. Lett.* 36, (1976) 623–626.

79. Hämäläinen, K., D.P. Siddons, J.B. Hastings, and L.E. Berman. Elimination of the inner-shell lifetime broadening in X-ray-absorption spectroscopy. *Phys. Rev. Lett.* 67, (1991) 2850–2853.

80. Bergmann, U., P. Glatzel, and S.P. Cramer. Bulk-sensitive XAS characterization of light elements: From X-ray Raman scattering to X-ray Raman spectroscopy. *Microchemical Journal* 71, (2002) 221–230.

81. Van Bokhoven, J.A. et al. Activation of oxygen on gold/alumina catalysts: *In situ* high-energy-resolution fluorescence and time-resolved X-ray spectroscopy. *Angew. Chem.* 118, (2006) 4767–4770.

82. Singh, J. et al. Generating highly active partially oxidized platinum during oxidation of carbon monoxide over Pt/Al 2O 3: *In situ*, time-resolved, and high-energy-resolution X-ray absorption spectroscopy. *Angew. Chem. Int. Ed.* 47, (2008) 9260–9264.

83. Kleymenov, E. et al. Structure of the methanol synthesis catalyst determined by *in situ* HERFD XAS and EXAFS. *Catal. Sci. Technol.* 2, (2012) 373.

84. Carra, P., M. Fabrizio, and B. Thole. High resolution X-ray resonant raman scattering. *Phys. Rev. Lett.* 74, (1995) 3700–3703.

85. Sarode, P.R. Effects of chemical combination on X-ray Kα emission spectra of chromium. *X-ray Spectrum.* 22, (1993) 138–144.

86. Baydaş, E. and E. Öz. Effects of chemical combination on X-ray Kα and Kβ 1,3 emission spectra of Co. *Phys Scripta* 81, (2010) 015302.

87. Yasuda, S. Chemical effects on the X-ray K emission spectra of phosphorus in organic compounds. *B. Chem. Soc. Jpn.* 57, (1984) 3122–3124.

88. Kawai, J. et al. Charge transfer effects on the chemical shift and the line width of the CuKα X-ray flourescence spectra of copper oxides. *Solid State Commun.* 70, (1989) 567–571.

89. Konishi, T. et al. Chemical shift and lineshape of high-resolution Ni Ka X-ray fluorescence spectra. *X-ray Spectrom.* 28, (1999) 470–477.

90. Sá, J. et al. The oxidation state of copper in bimetallic (Pt–Cu, Pd–Cu) catalysts during water denitration. *Catal. Sci. Technol.* 2, (2012) 794.

91. Friebel, D. et al. Balance of nanostructure and bimetallic interactions in Pt model fuel cell catalysts: *In situ* XAS and DFT study. *J. Am. Chem. Soc.* 134, (2012) 9664–9671.
92. Manyar, H.G. et al. High energy resolution fluorescence detection XANES—An *in situ* method to study the interaction of adsorbed molecules with metal catalysts in the liquid phase. *Catal. Sci. Technol.* 3, (2013) 1497.
93. Sparks, C.J. Inelastic resonance emission of X-rays: Anomalous scattering associated with anomalous dispersion. *Phys. Rev. Lett.* 33, (1974) 262.
94. Tulkki, J. and T. Åberg. Behavior of raman resonance scattering across the K-X-ray absorption-edge. *J. Phys. B-at Mol. Opt.* 15, (1982) L435–L440.
95. Tulkki, J. Evolution of the inelastic X-ray scattering by L and M electrons into K fluorescence in argon. *Phys. Rev. A* 27, (1983) 3375–3378.
96. Hayashi, H., Y. Udagawa, W.A. Caliebe, and C.C. Kao. Lifetime-broadening removed X-ray absorption near edge structure by resonant inelastic X-ray scattering spectroscopy. *Chem. Phys. Lett.* 371, (2003) 125–130.
97. Hayashi, H. et al. Lifetime-broadening-suppressed/free XANES spectroscopy by high-resolution resonant inelastic X-ray scattering. *Phys. Rev. B* 68, (2003) 045122.
98. Szlachetko, J. et al. High-resolution study of X-ray resonant raman scattering at the K edge of silicon. *Phys. Rev. Lett.* 97, (2006) 073001.
99. Szlachetko, J. et al. High-resolution study of the X-ray resonant Raman scattering process around the 1s absorption edge for aluminium, silicon, and their oxides. *Phys. Rev. A* 75, (2007) 022512.
100. Szlachetko, J. et al. High energy resolution off-resonant spectroscopy at sub-second time resolution: (Pt(acac) 2) decomposition. *Chemical Communications* 48, (2012) 10898–10900.
101. Kavčič, M. et al. Hard X-ray absorption spectroscopy for pulsed sources. *Phys. Rev. B* 87, (2013) 075106.
102. Friedbacher, G. and H. Bubert. *Surface and Thin Film Analysis: A Compendium of Principles, Instrumentation, and Applications*, 2nd ed. Weinheim, Germany: Wiley-VCH Verlag GmbH. doi: 10.1002/9783527636921.
103. P. Williams, Secondary ion mass spectrometry. *Ann. Rev. Mater. Sci.* 15, (1985) 517–548.
104. Sodhi, R.N.S. Time-of-flight secondary ion mass spectrometry (TOF–SIMS): Versatility in chemical and imaging surface analysis. *Analyst* 129, (2004) 483–487.
105. Chu, W. and J. Liu. Rutherford backscattering spectrometry: Reminiscences and progresses. *Mater. Chem. Phys.* 46 (2–3), (1996) 183–188.
106. Copel, M. Medium-energy ion scattering for analysis of microelectronic materials. *IBM J. Res. Dev.* 44 (4), (2000) 571–582.
107. Watts, J.F. and J. Wolstenholme. *An Introduction to Surface Analysis by XPS and AES.* Chichester, UK: John Wiley & Sons. doi: 10.1002/0470867930.
108. Tofterup, A.L. Theory of elastic and inelastic scattering of electrons emitted from solids: Energy spectra and depth profiling in XPS/AES. *Surf. Sci.* 167, (1986) (1).
109. Jellison Jr, G.E. Generalized ellipsometry for materials characterization. *Thin Solid Films* 450, (1) (2004) 42–50.
110. Vernon-Parry, K. Scanning electron microscopy: An introduction. *III-Vs Rev.* 13, (4) (2000) 40–44.
111. Binnig, G., H. Rohrer, C. Gerber, and E. Weibel. Surface studies by scanning tunneling microscopy. *Phys. Rev. Lett.* 49, (1982) 57–61.
112. Abou-Ras, D., R. Caballero, C.-H. Fischer et al. Comprehensive comparison of various techniques for the analysis of elemental distributions in thin films. *Microsc. Microanal.* 17, (2011) 728–751.
113. Escobar Galindo, R., R. Gago, D. Duday, and C. Palacio. Towards nanometric resolution in multilayer depth profiling: A comparative study of RBS, SIMS, XPS, and GDOES. *Anal. Bioanal. Chem.* 396, (2010) 2725–2740.

114. Baake, O., P. Hoffmann, S. Flege et al. Nondestructive characterization of nanoscale layered samples. *Anal. Bioanal. Chem.* 393, (2009) 623–634.
115. Johansson, S.A.E., J.L. Campbell, and K.G. Malmqvist. (1995) *Particle-Induced X-ray Emission Spectrometry (PIXE)*. New York: Wiley-Interscience.
116. Pahlke, S. Quo vadis total reflection X-ray fluorescence? *Spectrochim. Acta B* 58, (2003) 2025–2038.
117. Klockenkamper, R. *Total-Reflection X-ray Fluorescence Analysis*. (1995) New York: John Wiley & Sons.
118. de Bokx, P., C. Kok, A. Bailleul, G. Wiener, and H. Urbach. Grazing-emission X-ray fluorescence spectrometry; principles and applications. *Spectrochim. Acta B* 52, (1997) 829–840.
119. Compton, A.H. The total reflexion of X-rays. *Philosophical Magazine Series* 6, 45, (1923) 1121–1131.
120. Parratt, L.G. Surface studies of solids by total reflection of X-rays. *Phys. Rev.* 95, (1954) 359–369.
121. Yoneda Y. and T. Horiuchi. Optical flats for use in X-ray spectrochemical microanalysis. *Rev. Sci. Instrum.* 42, (1971) 1069.
122. Hannes A. and P. Wobrauschek. A method for quantitative X-ray fluorescence analysis in the nanogram region. *Nuclear Instruments and Methods* 114, (1974) 157–158.
123. Leenaers, A.J.G, J.J.A.M. Vrakking, and D.K.G de Boer. Glancing incidence X-ray analysis: More than just reflectivity! *Spectrochim. Acta B* 52, (1997) 805–812.
124. Stoev, K.N. and K. Sakurai. Recent theoretical models in grazing incidence X-ray reflectometry. *Rigaku Journal* 14, (1997) 22–37.
125. de Boer, D.K.G. X-ray scattering and X-ray fluorescence from materials with rough interfaces. *Phys. Rev. B* 53, (1996) 6048–6064.
126. Tsuji, K., T. Yamada, T. Utaka, and K. Hirokawa. The effects of surface roughness on the angle-dependent total-reflection X-ray fluorescence of ultrathin films. *J. Appl. Phys.* 78, (1995) 969–973.
127. Henke, B.L., E.M. Gullikson, and J.C. Davis. X-ray Interactions: Photoabsorption, scattering, transmission, and reflection at E = 50–30000 eV, Z = 1–92. *Atomic Data and Nuclear Data Tables* 54, (1993) 181–342.
128. Bedzyk, M.J., G.M. Bommarito, and J.S. Schildkraut. X-ray standing waves at a reflecting mirror surface. *Phys. Rev. Lett.* 62, (1989) 1376–1379.
129. von Bohlen, A., M. Kramer, C. Sternemann, and M. Paulus. The influence of X-ray coherence length on TXRF and XSW and the characterization of nanoparticles observed under grazing incidence of X-rays. *J. Anal. At. Spectrom.* 24, (2009) 792–800.
130. Meirer, F., A. Singh, P. Pianetta et al. Synchrotron radiation-induced total reflection X-ray fluorescence analysis. *Anal. Chem.* 29, (2010) 479–496.
131. Brücher, M., A. von Bohlen, and R. Hergenröder. The distribution of the contrast of X-ray standing waves fields in different media. *Spectrochim. Acta B* 71–72, (2012) 62–69.
132. de Boer, D.K.G. Glancing-incidence X-ray fluorescence of layered materials. *Phys. Rev. B* 44, (1991) 498–511.
133. Krämer, M., A. von Bohlen, C. Sternemann, M. Paulus, and R. Hergenröder. Synchrotron radiation induced X-ray standing waves analysis of layered structures. *Appl. Surf. Sci.* 253, (2007) 3533–3542.
134. Windt, D.L. IMD—Software for modeling the optical properties of multilayer films. *Comput. Phys.* 12, (1998) 360–370.

135. Tiwari, M.K., G.S. Lodha, and K.J.S. Sawhney. Applications of the "CATGIXRF" computer program to the grazing incidence X-ray fluorescence and X-ray reflectivity characterization of thin films and surfaces. *X-ray Spectrom.* 39, (2010) 127–134.
136. Batterman, B.W. Detection of foreign atom sites by their X-ray fluorescence scattering. *Phys. Rev. Lett.* 22, (1969) 703–705.
137. Cowan, P.L., J.A. Golovchenko, and M.F. Robbins. X-ray standing waves at crystal surfaces. *Phys. Rev. Lett.* 44, (1980) 1680–1683.
138. Patel, J.R. and J.A. Golovchenko. X-ray-standing-wave atom location in heteropolar crystals and the problem of extinction. *Phys. Rev. Lett.* 50, (1983) 1858–1861.
139. Misram, N.L. and K.D.S. Mudher. Total reflection X-ray fluorescence: A technique for trace element analysis in materials. *Prog. Cryst. Growth Ch.* 45, (2002) 65–74.
140. Streli, C., G. Pepponi, P. Wobrauschek et al. Analysis of low Z elements on Si wafer surfaces with synchrotron radiation induced total reflection X-ray fluorescence at SSRL, Beamline 3-3: Comparison of droplets with spin coated wafers. *Spectrochim. Acta B* 58, (2003) 2105–2112.
141. von Bohlen, A. Total reflection X-ray fluorescence and grazing incidence X-ray spectrometry—Tools for micro- and surface analysis. A review. *Spectrochim. Acta B* 64, (9), (2009) 821–832.
142. Egorov, V., E. Egorov, and M.M. Afanas'ev. Comparative analysis of TXRF-spectroscopy efficiency under the testing target excitation by fluxes formed by the slit-cut system and waveguide-resonator. *J. Surf. Investig.-X-ray Synchro.* 2, (2008) 904–912.
143. Yang, J., D. Zhao, Q. Xu, and X. Ding. Development and application of glancing incident X-ray fluorescence spectrometry using parallel polycapillary X-ray lens. *Appl. Surf. Sci.* 255, (2009) 6439–6442.
144. Nakano, K., K. Tanaka, X. Ding, and K. Tsuji. Development of a new total reflection X-ray fluorescence instrument using polycapillary X-ray lens. *Spectrochim. Acta B* 61, (2006) 1105–1109.
145. Alov, N.V. Total reflection X-ray fluorescence analysis: Physical foundations and analytical application (A review). *Inorg. Mater.* 47, (2011) 1487–1499.
146. Wobrauschek, P. Total reflection X-ray fluorescence analysis: A review. *X-ray Spectrom.* 36, (2007) 289–300.
147. Stoev, K.N. and K. Sakurai. Review on grazing incidence X-ray spectrometry and reflectometry. *Spectrochim. Acta B* 54, (1999) 41–82.
148. Klockenkämper, R. Challenges of total reflection X-ray fluorescence for surface- and thin-layer analysis. *Spectrochim. Acta Part B* 61, (2006) 1082–1090.
149. Beckhoff, B., R. Fliegauf, M. Kolbe, M. Müller, J. Weser, and G. Ulm. Reference-free total reflection X-ray fluorescence analysis of semiconductor surfaces with synchrotron radiation. *Anal. Chem.* 79, (2007) 7873–7882.
150. Shin, N.S., C.H. Chang, Y.M. Koo, and H. Padmore. Synchrotron radiation excited total reflection X-ray fluorescence quantitative analysis of Si wafer by absolute fluorescence intensity calculation. *Matter. Lett.* 49, (2001) 38–42.
151. Kubala-Kukuś, A., D. Banaś, M. Pajek et al. Synchrotron radiation based micro X-ray fluorescence analysis of the calibration samples used in surface sensitive total reflection and grazing emission X-ray fluorescence techniques. *Rad. Phys. Chem.* 93, (2012) 117–122.
152. Klockenkämper, R. and A. von Bohlen. Determination of the critical thickness and the sensitivity for thin-film analysis by total reflection X-ray fluorescence spectrometry. *Spectrochim. Acta B* 44, (1989) 461–469.
153. Klockenkämper, R. and A. von Bohlen. Total-reflection X-ray fluorescence moving towards nanoanalysis: A survey. *Spectrochim. Acta B* 56, (2001) 2005–2018.

154. Borgese, L., A. Zacco, E. Bontempi et al. Total reflection of X-ray fluorescence (TXRF): A mature technique for environmental chemical nanoscale metrology. *Meas. Sci. Technol.* 20, (2009) 084027.

155. Antosz, F.J., Y. Xiang, A.R. Diaz, A.J. Jensen. The use of total reflectance X-ray fluorescence (TXRF) for the determination of metals in the pharmaceutical industry. *J. Pharmaceut. Biomed.* 62, (2012) 17–22.

156. Szoboszlai, N., Z. Polgári, V.G. Mihucz, and G. Záray. Recent trends in total reflection X-ray fluorescence spectrometry for biological applications. *Anal. Chim. Acta* 633, (2009) 1–18.

157. Cinosi, A., N. Andriollo, and G.P.D. Monticelli. A novel total reflection X-ray fluorescence procedure for the direct determination of trace elements in petrochemical products. *Anal. Bioanal. Chem.* 399, (2011) 927–933.

158. Hellin, D., S.D. Gendt, N. Valckx, P.W. Mertens, and C. Vinckier. Trends in total reflection X-ray fluorescence spectrometry for metallic contamination control in semiconductor nanotechnology. *Spectrochim. Acta B* 61, (2006) 496–514.

159. Baur, K., S. Brennan, P. Pianetta, and R. Opila. Looking at trace impurities on silicon wafers with synchrotron radiation. *Anal. Chem.* 74, (2002) 608 A–616 A.

160. Klockenkämper, R. Challenges of total reflection X-ray fluorescence for surface- and thin-layer analysis. *Spectrochim. Acta B* 61, (2006) 1082–1090.

161. Mori, Y., K. Uemura, H. Kohno, M. Yamagami, and Y. Iizuka. Sweeping-TXRF: A nondestructive technique for the entire surface characterization of metal contaminations on semiconductor wafers. *IEEE T. Semiconduct. M.* 18, (2005) 569–574.

162. Mori, Y., K. Uemura, and K. Shimanoe. A depth profile fitting model for a commercial total reflection X-ray fluorescence spectrometer. *Spectrochim. Acta B* 52, (1997) 823–828.

163. Li, W., J. Zhu, X. Ma et al. Geometrical factor correction in grazing incident X-ray fluorescence experiment. *Rev. Sci. Instrum.* 83, (2012) 053114.

164. Unterumsberger, R., B. Pollakowski, M. Müller, and B. Beckhoff. Complementary characterization of buried nanolayers by quantitative X-ray fluorescence spectrometry under conventional and grazing incidence conditions. *Anal. Chem.* 83, (2011) 8623–8628.

165. Makhotkin, I.A., E. Louis, R.W.E. van de Kruijs et al. Determination of the density of ultrathin La films in La/B$_4$C layered structures using X-ray standing waves. *Phys. Status Solidi A* 208, (2011) 2597–2600.

166. Awaji, N. Wavelength dispersive grazing incidence X-ray fluorescence of multilayer thin films. *Spectrochim. Acta B* 59, (2004) 1133–1139.

167. Streeck, C., B. Beckhoff, F. Reinhardt et al. Elemental depth profiling of Cu(In,Ga) Se$_2$ thin films by reference-free grazing incidence X-ray fluorescence analysis. *Nucl. Instrum. Meth. B* 268, (2010) 277–281.

168. Hönicke, P., B. Beckhoff, M. Kolbe, S. List, T. Conard, and H. Struyff. Depth-profiling of vertical sidewall nanolayers on structured wafers by grazing incidence X-ray fluorescence. *Spectrochim. Acta B* 63, (2008) 1359–1364.

169. Sánchez, H.J. and C.A. Pérez. Study of copper surface oxidation by grazing angle X-ray excitation. *Spectrochim. Acta B* 65, (2010) 466–470.

170. Sánchez, H.J., C.A. Pérez, R.D. Pérez, and M. Rubio. Surface analysis by total-reflection X-ray fluorescence. *Rad. Phys. Chem.* 48, (1996) 325–331.

171. Weisbrod, U., R. Gutschke, J. Knoth, and H. Schwenke. Total reflection X-ray fluorescence spectrometry for quantitative surface and layer analysis. *Appl. Phys. A: Mater.* 53, (1991) 449–45.

172. Zheludeva, S., N. Novikova, N. Stepina, E. Yurieva, and O. Konovalov. Molecular organization in protein-lipid film on the water surface studied by X-ray standing wave measurements under total external reflection. *Spectrochim. Acta B* 63, (2008) 1399–1403.

173. Guico, R.S., S. Narayanan, J. Wang, and K.R. Shull. Dynamics of polymer/metal nano-composite films at short times as studied by X-ray standing waves. *Macromolecules* 37, (2004) 8357–8363.
174. Wang, J., M. Caffrey, M.J. Bedzyk, and T.L. Penner. *Langmuir* 17, (2001) 3671–3681.
175. de Carvalho, H.W.P., A.P.L. Batista, T.C. Ramalho, C.A. Pérez, and A.L. Gobbi. The interaction between atoms of Au and Cu with clean Si(111) surface: A study combining synchrotron radiation grazing incidence X-ray fluorescence analysis and theoretical calculations. *Spectrochim. Acta A* 74, (2009) 292–296.
176. Reinhardt, F., J. Osán, S. Török, A.E. Pap, M. Kolbe, and B. Beckhoff. Reference-free quantification of particle-like surface contaminations by grazing incidence X-ray fluorescence analysis. *J. Anal. At. Spectrom.* 27, (2012) 248–255.
177. Tiwari, M.K., K.J.S. Sawhney, T.-L. Lee, S.G. Alcock, and G.S. Lodha. Probing the average size of self-assembled metal nanoparticles using X-ray standing waves. *Phys. Rev. B* 80, (2009) 035434.
178. Von Bohlen, A., M. Brücher, B. Holland, R. Wagner, and R. Hergenröder. X-ray standing waves and scanning electron microscopy—Energy dispersive X-ray emission spectroscopy study of gold nanoparticles. *Spectrochim. Acta B* 65, (2010) 409–414.
179. Kitts, K., Y. Choi, P.J. Eng, S.K. Ghose, S.R. Sutton, and B. Rout. Application of grazing incidence X-ray fluorescence technique to discriminate and quantify implanted solar wind. *J. Appl. Phys.* 105, (2009) 064905.
180. Hönicke, P., B. Beckhoff, M. Kolbe, D. Giubertoni, J. van den Berg, and G. Pepponi. Depth profile characterization of ultra shallow junction implants. *Anal. Bioanal. Chem.* 396, (2010) 2825–2832.
181. Pepponi, G., D. Giubertoni, M. Bersani et al. Grazing incidence X-ray fluorescence and secondary ion mass spectrometry combined approach for the characterization of ultra-shallow arsenic distribution in silicon. *J. Vac. Sci. Technol. B* 28, (2010) C1C59–C1C64.
182. Becker, R.S., J.A. Golovchenko, and J.R. Patel. X-ray evanescent-wave absorption and emission. *Phys. Rev. Lett.* 50, (1983) 153–156.
183. Potton, R.J. Reciprocity in optics. *Reports on Progress in Physics* 67, (2004) 717–754.
184. Noma, T., A. Iida, K. Sakurai. Fluorescent-X-ray-interference effect in layered materials. *Phys. Rev. B* 48, (1993) 17524–17526.
185. Sasaki, Y.C., Y. Suzuki, Y. Tomioka, and A. Fukuhara. Observation of an interference effect for fluorescent X-rays. *Phys. Rev. B* 48, (1993) 7724–7726.
186. Pérez, R.D., H. Sánchez, and M. Rubio. Efficient calculation method for glancing angle X-ray techniques. *X-ray Spectrom.* 31, (2002) 296–299.
187. Urbach, H.P. and P.K. de Bokx. Grazing emission X-ray fluorescence from multilayers. *Phys. Rev. B* 63, (2001) 085408.
188. Bekshaev, A., J. de Hoog, and R. van Grieken. Grazing-emission electron probe micro-analysis of particles near the substrate edge. *Spectrochim. Acta B* 56, (2001) 2385–2395.
189. Kayser, Y., J. Szlachetko, D. Banaś et al. High-energy-resolution grazing emission X-ray fluorescence applied to the characterization of thin Al films on Siv. *Spectrochim. Acta B* 88, (2013) 136–149.
190. Yang, J., K. Tsuji, X. Lin, D. Han, and X. Ding. A micro X-ray fluorescence analysis method using polycapillary X-ray optics and grazing exit geometry. *Thin Solid Films* 517, (2009) 3357–3361.
191. Kuczumow, A., M. Schmeling, and R. van Grieken. Critical assessment and proposal for reconstruction of a grazing emission X-ray fluorescence instrument. *J. Anal. At. Spectrom.* 15, (2000) 535–542.
192. Pérez, R. and H. Sánchez. New spectrometer for grazing exit X-ray fluorescence. *Rev. Sci. Instrum.* 68, (1997) 2681–2684.

193. de Bokx, P.K. and H.P. Urbach. Laboratory grazing-emission X-ray fluorescence spectrometer. *Rev. Sci. Instrum.* 66, (1995) 15–19.
194. Claes, M., P. de Bokx, and R. van Grieken. Progress in laboratory grazing emission X-ray fluorescence spectrometry. *X-ray Spectrom.* 28, (1999) 224–229.
195. Hasegawa, S., S. Ino, Y. Yamamoto, and H. Daimon. Chemical analysis of surfaces by total-reflection angle X-ray spectroscopy in RHEED experiments (RHEED-TRAXS). *Jpn. J. Appl. Phys.* 24, (1985) L387–L390.
196. Sasaki Y.C. and K. Hirokawa. New nondestructive depth profile measurement by using a refracted X-ray fluorescence method. *Appl. Phys. Lett.* 58, (1991) 1384–1386.
197. Claes, M., K.V. Dyck, H. Deelstra, and R. van Grieken. Determination of silicon in organic matrices with grazing-emission X-ray fluorescence spectrometry. *Spectrochim. Acta B* 54, (1999) 1517–1524.
198. Spolnik, Z.M., M. Claes, and R. van Grieken. Determination of trace elements in organic matrices by grazing-emission X-ray fluorescence spectrometry. *Anal. Chim. Acta* 401, (1999) 293–298.
199. Kubala-Kukuś, A., D. Banaś, W. Cao et al. Observation of ultralow-level Al impurities on a silicon surface by high-resolution grazing emission X-ray fluorescence excited by synchrotron radiation. *Phys. Rev. B* 80, (2009) 113305.
200. de Gendt, S., K. Kenis, M. Baeyens et al. Silicon surface metal contamination measurements using grazing-emission XRF spectrometry. *Mater. Res. Soc. Symp. Proc.* 477, (1997) 397–402.
201. Holynska, B., M. Olko, B. Ostachowicz et al. Performance of total reflection and grazing emission X-ray fluorescence spectrometry for the determination of trace metals in drinking water in relation to other analytical techniques. *Fresen. J. Anal. Chem.* 362, (1998) 294–298.
202. Claes, M., R. van Ham, K. Janssens, R. van Grieken, R. Klockenkämper, and A. von Bohlen. Micro-analysis of artists' pigments by grazing-emission X-ray fluorescence spectrometry. *Adv. X-ray Anal.* 41, (1998) 262–277.
203. Blockhuys, F., M. Claes, R. van Grieken, and H.J. Geise. Assessing the molecular weight of a conducting polymer by grazing emission XRF. *Anal. Chem.* 72, (2000) 3366–3368.
204. Szlachetko, J., D. Banaś, A. Kubala-Kukuś et al. Application of the high-resolution grazing-emission X-ray fluorescence method for impurities control in semiconductor nanotechnology. *J. Appl. Phys.* 105, (2009) 086101.
205. Kuczumow, A., M. Claes, M. Schmeling, R. van Grieken, and S. de Gendt. Quantification problems in light element determination by grazing emission X-ray fluorescence. *J. Anal. At. Spectrom.* 15, (2000) 415–421.
206. Spolnik, Z.M., M. Claes, R.E. van Grieken, P.K. de Bokx, and H.P. Urbach. Quantification in grazing emission X-ray fluorescence spectrometry. *Spectrochim. Acta B* 54, (1999) 1525–1537.
207. Claes, M., P. de Bokx, N. Willard, P. Veny, and R. van Grieken. Optimization of sample preparation for grazing emission X-ray fluorescence in micro- and trace analysis applications. *Spectrochim. Acta B* 52, (1997) 1063–1070.
208. Bekshaev, A. and R. van Grieken. Substrates with a periodic surface structure in grazing-exit X-ray microanalysis. *Spectrochim. Acta B* 57, (2002) 865–882.
209. Noma, T., K. Takada, and A. Iida. Surface-sensitive X-ray fluorescence and diffraction analysis with grazing-exit geometry. *X-ray Spectrom.* 28, (1999) 433–439.
210. Noma, T., A. Iida, and K. Sakurai. Fluorescent-X-ray-interference effect in layered materials. *Phys. Rev. B* 48, (1993) 17524–17526.
211. Noma, T., H. Miyata, and S. Ino. Grazing exit X-ray fluorescence spectroscopy for thin-film analysis. *Jpn. J. Appl. Phys.* 31, (1992) L900–L903.

212. Skytt, P., B. Gålnander, T. Nyberg, J. Nordgren, and P. Isberg, Probe depth variation in grazing exit soft-X-ray emission spectroscopy. *Nucl. Instrum. Meth. A* 384, (1997) 558–562.
213. Monaghan, M.L., T. Nigam, M. Houssa, S.D. Gendt, H.P. Urbach, and P.K. de Bokx. Characterization of silicon oxynitride films by grazing-emission X-ray fluorescence spectrometry. *Thin Solid Films* 359, (2000) 197–202.
214. Wiener, G., S.J. Kidd, C.A.H. Mutsaers, R.A.M. Wolters, and P.K. de Bokx. Characterization of titanium nitride layers by grazing-emission X-ray fluorescence spectrometry. *Appl. Surf. Sci.* 125, (1998) 129–136.
215. Tsuji, K., H. Takenaka, K. Wagatsuma, P.K. de Bokx, and R. van Grieken. Enhancement of X-ray fluorescence intensity from an ultra-thin sandwiched layer at grazing-emission angles. *Spectrochim. Acta B* 54, (1999) 1881–1888.
216. Koshelev, I., A. Paulikas, M. Beno et al. Chromium-oxide growth on Fe–Ni–Cr alloy studied with grazing-emission X-ray fluorescence. *Oxid. Met.* 68, (2007) 37–51.
217. Kayser, Y., D. Banaś, W. Cao et al. Depth profiling of dopants implanted in Si using the synchrotron radiation based high-resolution grazing emission technique. *X-ray Spectrom.* 41, (2012) 98–104.
218. Meirer, F., G. Pepponi, C. Streli, P. Wobrauschek, and N. Zoeger. Grazing exit versus grazing incidence geometry for X-ray absorption near edge structure analysis of arsenic traces. *J. Appl. Phys.* 105, (2009) 074906.
219. Emoto, T., Y. Sato, Y. Konishi, X. Ding, and K. Tsuji. Development and applications of grazing exit micro X-ray fluorescence instrument using a polycapillary X-ray lens. *Spectrochim. Acta B* 59, (2004) 1291–1294.
220. Tsuji, K. and F. Delalieux. Feasibility study of three-dimensional XRF spectrometry using μ-X-ray beams under grazing-exit conditions. *Spectrochim. Acta B* 58, (2003) 2233–2238.
221. Noma, T. and A. Iida. Surface analysis of layered thin films using a synchrotron x-ray microbeam combined with a grazing-exit condition. *Rev. Sci. Instrum.* 65, (1994) 837.
222. Tsuji, K. Grazing-exit electron probe X-ray microanalysis (GE-EPMA): Fundamental and applications. *Spectrochim. Acta B* 60, (2005) 1381–1391.
223. Spolnik, Z., J. Zhang, K. Wagatsuma, and K. Tsuji. Grazing-exit electron probe X-ray microanalysis of ultra-thin films and single particles with high-angle resolution. *Anal. Chim. Acta*, 455, (2002) 245–252.
224. Bekshaev, A. and R. van Grieken. Interference technique in grazing-emission electron probe microanalysis of submicrometer particles. *Spectrochim. Acta B* 56, (2001) 503–515.
225. Lennard, W.N., J.K. Kim, and L. Rodríguez-Fernández. Surface sensitive particle-induced X-ray emission. *Nucl. Instrum. Meth. B* 189, (2002) 49–55.
226. Tsuji, K., M. Huisman, Z. Spolnik et al. Comparison of grazing-exit particle-induced X-ray emission with other related methods. *Spectrochim. Acta B* 55, (2000) 1009–1016.
227. Schwenke, H., J. Knoth, P.A. Beaven, R. Kiehn, and J. Buhrz. A laser plasma X-ray source for the analysis of wafer surfaces by grazing emission X-ray fluorescence spectrometry. *Spectrochim. Acta B* 59, (2004) 1159–1164.
228. Sánchez, H.J. Direct comparison of total reflection techniques by using a plate beam-guide. *X-ray Spectrom.* 31, (2002) 145–149.
229. Sánchez, H.J. Detection limit calculations for the total reflection techniques of X-ray fluorescence analysis. *Spectrochim. Acta B* 56, (2001) 2027–2036.
230. Hönicke, P., Y. Kayser, B. Beckhoff et al. Characterization of ultra-shallow aluminum implants in silicon by grazing incidence and grazing emission X-ray fluorescence spectroscopy. *J. Anal. At. Spectrom.* 27, (2012) 1432–1438.
231. Kavcic, M., M. Zitnik, K. Bucar, and J. Szlachetko. Application of wavelength dispersive X-ray spectroscopy to improve detection limits in X-ray analysis. *X-ray Spectrom.* 40, (2011) 2–6.

232. Pahlke, S., L. Fabry, L. Kotz, C. Mantler, and T. Ehmann. Determination of ultra trace contaminants on silicon wafer surfaces using total-reflection X-ray fluorescence TXRF "state-of-the-art." *Spectrochim. Acta B* 56, (2001) 2261–2274.

233. Tsuji, K., K. Wagatsuma, and T. Oku. Glancing-incidence and glancing-takeoff X-ray fluorescence analysis of Ni–GaAs interface reactions. *X-ray Spectrom.* 29, (2000) 155–160.

234. Tsuji, K., S. Sato, K. Hirokawa. Glancing-incidence and glancing-takeoff X-ray fluorescence analysis of a Mn ultrathin film on an Au layer. *Thin Solid Films* 274, (1996) 18–22.

235. Kayser, Y., J. Szlachetko, and J. Sà. Scanning-free GEXRF by means of an angular dispersive arrangement with a two-dimensional position-sensitive area detector. *Rev. Sci. Instrum.* 84, (2013) 123102.

236. Shinoda, K., S. Sato, S. Suzuki et al. Nondestructive depth resolved analysis by using grazing exit fluorescence yield X-ray absorption spectroscopy. *J. Surf. Anal.* 15, (2009) 295–298.

237. Shinoda, K., S. Suzuki, K. Yashiro et al. Nondestructive depth-resolved chemical state analysis of (La, Sr) MnO$_3$ film under high temperature. *Surf. Interface Anal.* 42, (2010) 1650–1654.

238. Holfelder, I., B. Beckhoff, R. Fliegauf et al. Complementary methodologies for thin film characterization in one tool—A novel instrument for 450 mm wafers. *J. Anal. At. Spectrom.* 28, (2013) 549–557.

239. Jiang, H., J. Zhu, J. Xu, X. Wang, Z. Wang, and M. Watanabe. Determination of layer-thickness variation in periodic multilayer by X-ray reflectivity. *J. Appl. Phys.* 107, (2010) 103523.

240. Kolbe, M., B. Beckhoff, M. Krumrey, and G. Ulm. Comparison of reference-free X-ray fluorescence analysis and X-ray reflectometry for thickness determination in the nanometer range. *Appl. Surf. Sci.* 252, (2005) 49–52.

241. Krumrey, M., M. Hoffmann, G. Ulm, K. Hasche, P. Thomsen-Schmidt. Thickness determination for SiO2 films on Si by X-ray reflectometry at the Si K edge. *Thin Solid Films* 459, (2004) 241–244.

242. Zargham, A., T. Schmidt, J.I. Flege et al. On revealing the vertical structure of nanoparticle films with elemental resolution: A total external reflection X-ray standing waves study. *Nucl. Instrum. Meth. B* 268, (2010) 325–328.

243. Rajput, P., A. Gupta, S. Rajagopalan, and A.K. Tyagi. Fe diffusion in amorphous Si studied using X-ray standing wave technique. *AIP Advances* 2, (2012) 012159.

244. Li, W., J. Zhu, H. Li et al. Ni layer thickness dependence of the interface structures for Ti/Ni/Ti trilayer studied by X-ray standing waves. *Appl. Mater. Interfaces* 5, (2013) 404–409.

245. Naudon, A., J. Chihab, P. Goudeau, and J. Mimault. New apparatus for grazing X-ray reflectometry in the angle-resolved dispersive mode. *J. Appl. Cryst.* 22, (1989) 460–464.

246. Kawai, J. Total reflection X-ray photoelectron spectroscopy: A review. *J. Electron Spectrosc.* 178–179, (2010) 268–272.

247. Becker, C., M. Pagels, C. Zachaus et al. Chemical speciation at buried interfaces in high-temperature processed polycrystalline silicon thin-film solar cells on ZnO: Al. *J. Appl. Phys.* 113, (2013) 044519.

248. Pagels, M., F. Reinhardt, B. Pollakowski et al. GIXRF–NEXAFS investigations on buried ZnO/Si interfaces: A first insight in changes of chemical states due to annealing of the specimen. *Nucl. Instrum. Meth. B* 268, (2010) 370–373.

249. Pollakowski, B., P. Hoffmann, M. Kosinova et al. Nondestructive and nonpreparative chemical nanometrology of internal material interfaces at tunable high information depths. *Anal. Chem.* 85, (2013) 193–200.

250. Maurizio, C., M. Rovezzi, F. Bardelli, H.G. Pais, and F. D'Acapito. Setup for optimized grazing incidence X-ray absorption experiments on thin films on substrates. *Rev. Sci. Instrum.* 80, (2009) 063904.

251. Waychunas, G.A. Grazing-incidence X-ray absorption and emission spectroscopy. *Rev. Mineral. Geochem.* 49, (2002) 267–315.

252. Pollakowski, B., B. Beckhoff, F. Reinhardt, S. Braun, and P. Gawlitza. Speciation of deeply buried TiOx nanolayers with grazing-incidence X-ray fluorescence combined with a near-edge X-ray absorption fine structure investigation. *Phys. Rev. B* 77, (2008) 235408.

253. Suzuki, Y. Surface extended X-ray-absorption fine-structure spectroscopy measurement using the evanescent-wave effect of fluorescent X-ray. *Phys. Rev. B* 39, (1989) 3393–3395.

254. Takamatsu, D., T. Nakatsutsumi, S. Mori et al. Nanoscale observation of the electronic and local structures of LiCoO2 thin film electrode by depth-resolved X-ray absorption spectroscopy. *J. Phys. Chem. Lett.* 2, (2011) 2511–2514.

255. Beye, M., Ph. Wernet, C. Schüßler-Langeheine, and A. Föhlisch. Time resolved resonant inelastic X-ray scattering: A supreme tool to understand dynamics in solids and molecules. *J. Electron Spectrosc.* 188, (2013) 172–182.

256. Szlachetko, J., J. Sá, O.V. Safonova et al. *In situ* hard X-ray quick RIXS to probe dynamic changes in the electronic structure of functional materials. *J. Electron Spectrosc.* 188, (2013) 161–165.

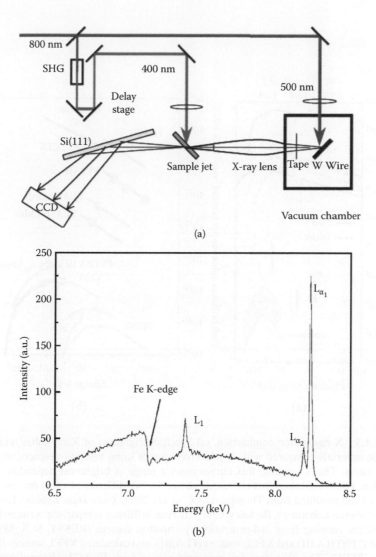

(a)

(b)

**FIGURE 1.3** (a) Schematic of a laser plasma X-ray source using a tungsten wire as the target, a microcapillary optic to focus the X-rays, and a dispersive detection scheme. (b) The X-ray spectrum generated by this source measured after transmission through a Fe-containing sample, showing the X-ray continuum and tungsten emission lines. (Reproduced from Chen, J. et al., 2007, *Journal of Physical Chemistry A*, 111, 38, 9326–9335.[24] With permission.)

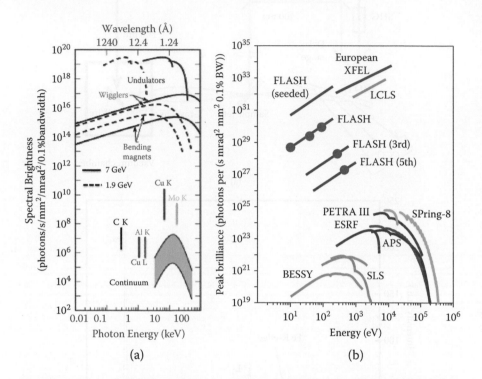

**FIGURE 1.5** X-ray source comparison. (a) Spectral brightness of X-ray tubes with various anode materials compared with synchrotron sources using bending magnets, wigglers, and undulators. The X-ray tube flux curves cover a range of brightness depending on the tube design. The synchrotron flux curves are shown for two different electron energies. (Plot reproduced and modified from Thompson, A.C. et al., 2009, *X-ray Data Booklet*, Lawrence Berkeley National Laboratory, Berkeley, CA.[10]) (b) Peak brilliance comparison between facility X-ray sources, ranging from 3rd-generation synchrotron sources (BESSY, SLS, SPring-8, APS, ESRF, PETRA III) and XFEL sources (FLASH), and calculated XFEL sources (LCLS, European XFEL). The blue dots are measured values from FLASH. (Reproduced from Ackermann, W. et al., 2007, *Nature Photonics*, 1, 6, 336–342.[49] With permission.)

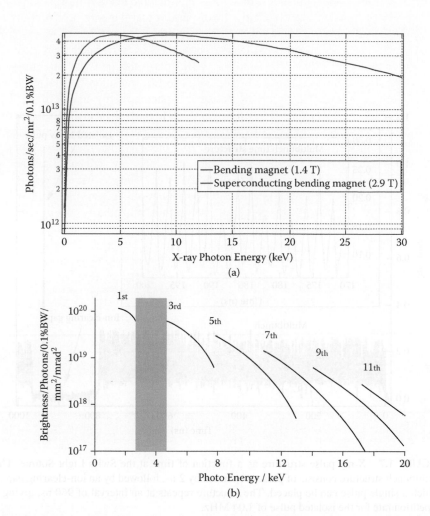

**FIGURE 1.6** (a) X-ray spectra generated by a bending magnet (red curve) and a superconducting bending magnet (blue curve) at the Swiss Light Source synchrotron. Flux curves calculated using XOP. (b) Calculated X-ray spectrum showing the harmonics of an in-vacuum, minigap undulator used at the microXAS beamline at the Swiss Light Source. The gray bar shows a gap in the X-ray spectrum where no X-ray photons are generated.

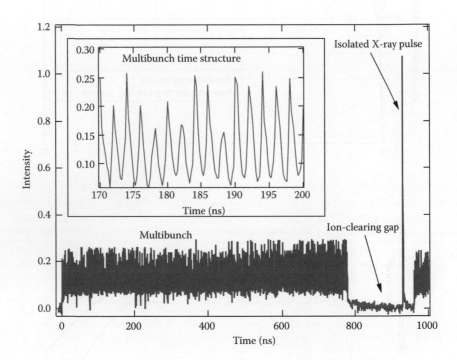

**FIGURE 1.7** X-ray pulse structure as a function of time at the Swiss Light Source. The multibunch structure consists of pulses separated by 2 ns, followed by an ion-clearing gap in which a single pulse can be placed. The structure repeats at an interval of 960 ns, giving a repetition rate for the isolated pulse of 1.04 MHz.

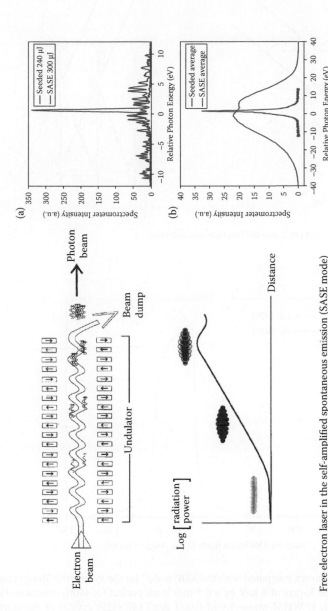

**FIGURE 1.8** Left: The formation of the microbunch structure in the electron pulse as it propagates through the undulators in an XFEL. (Reproduced from Sonntag, B., 2001, *Nuclear Instruments and Methods in Physics Research Section A—Accelerators Spectrometers Detectors and Associated Equipment,* 467–468, 8–15.[79] With permission.) Right: (a) Single-pulse spectra emitted from the LCLS in both unseeded (red curve) and self-seeded (blue curve) modes. (b) Average spectra for LCLS operating in unseeded (red curve) and self-seeded (blue curve) modes. Note the significant spectral narrowing when operating in the self-seeded mode. (Reproduced from Amann, J. et al., 2012, *Nature Photonics,* 6, 10, 693–698.[80] With permission.)

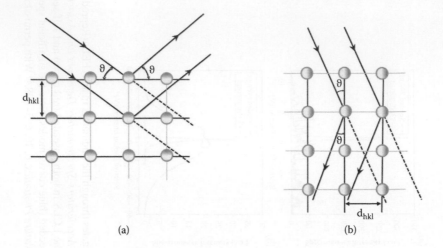

(a)　　　　　　　　　　　　(b)

**FIGURE 2.1**　(a) Bragg and (b) Laue diffraction geometries.

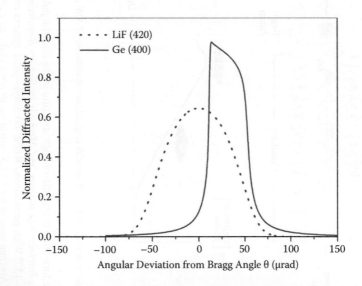

**FIGURE 2.2**　Rocking curves computed with the XOP code[13] for the symmetric Bragg case diffraction of σ-polarized X-rays of 8 keV by a 0.5 mm thick perfect Ge (400) crystal (solid blue curve) and a mosaic (FWHM mosaicity of 0.003 deg.) LiF (420) crystal of the same thickness (dotted red curve). For Ge, as a result of the refraction occurring at the front surface of the crystal, the centroid of the rocking curve is shifted by about 30 μrad from the Bragg angle.

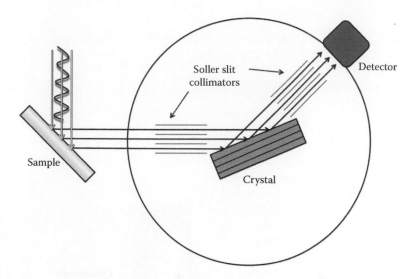

**FIGURE 2.3** Plane-crystal spectrometer geometry with Soller slit collimators.

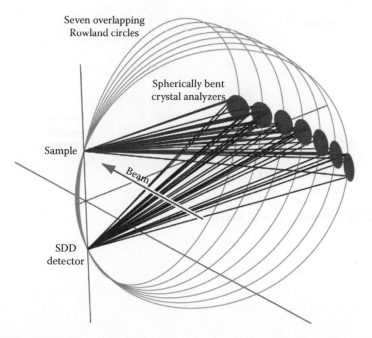

**FIGURE 2.7** Overlapping Rowland circles concept for a multicrystal Johann-type spectrometer. (Reprinted from Sokaras, D. et al., 2013, *Rev. Sci. Instrum.*, 84, 053102.[39] Copyright 2013, AIP Publishing LLC. With permission.)

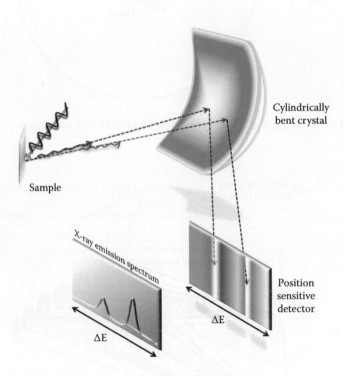

**FIGURE 2.8** Schematic view of von Hamos geometry.

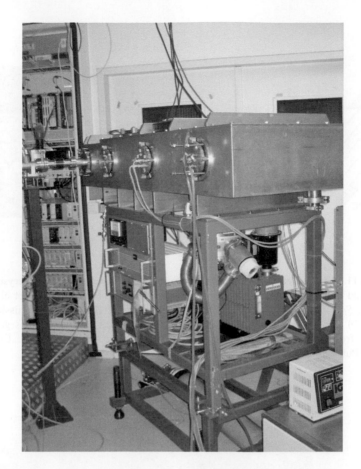

**FIGURE 2.9** The von Hamos X-ray spectrometer from the University of Fribourg installed at the PHOENIX beam line, Swiss Light Source (SLS), Villigen, Switzerland.

(a)                                                              (b)

**FIGURE 2.10** Crystal lamina permanently glued to an aluminum block machined to a precise concave cylindrical surface of a nominal radius of 25.4 cm (a), and the CCD 2D detector (b).

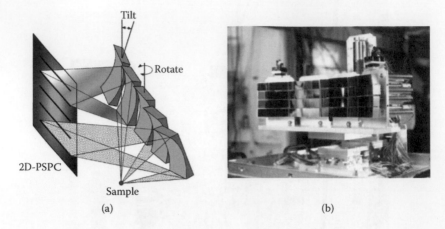

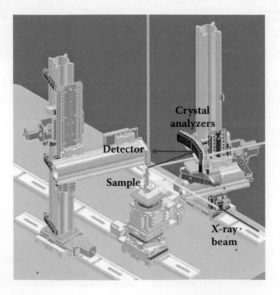

**FIGURE 2.11** Multiple crystals in von Hamos geometry. (a) Schematic of a five-crystal spectrometer and a 2D position-sensitive detector. (Reprinted from Hayashi, H. et al., 2004, *J. Electron Spectrosc. Relat. Phenom.*, 136, 191.[67] Copyright 2004. With permission from Elsevier.) (b) A picture of the 4 × 4 array of analyzer crystals in von Hamos geometry. (Reprinted from Alonso-Mori, R. et al., 2012, *Rev. Sci. Instrum.* 83, 073114.[69] Copyright 2012, AIP Publishing LLC. With permission.)

**FIGURE 2.12** Schematic drawing of the von Hamos spectrometer developed at the SLS SuperXAS beam line. (Reprinted from Szlachetko, J. et al., 2013, *J. Electron Spectrosc. Rel. Phenom.*, 188, 161.[72] Copyright 2004. With permission from Elsevier.)

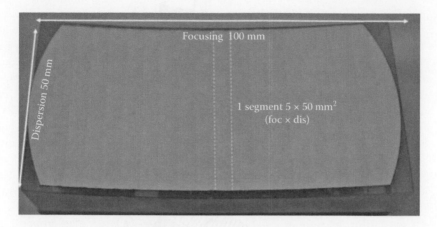

**FIGURE 2.13** A segmented Si(111) crystal bent to radius of 25 cm. (Reprinted from Szlachetko, J. et al., 2012, *Rev. Sci. Instrum.* 83, 103105.[71] With permission. Copyright 2012, AIP Publishing LLC.)

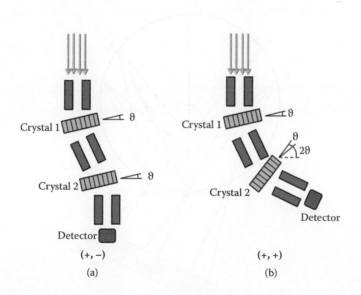

**FIGURE 2.14** (a) Parallel nondispersive (+,-) and (b) antiparallel dispersive (+,+) setups of a Laue double flat crystal spectrometer.

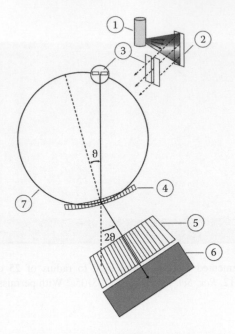

**FIGURE 2.16** Schematic drawing of the modified DuMond slit geometry: (1) X-ray tube, (2) target, (3) slit, (4) cylindrically bent crystal, (5) Soller slit collimator, (6) detector, and (7) focal circle.

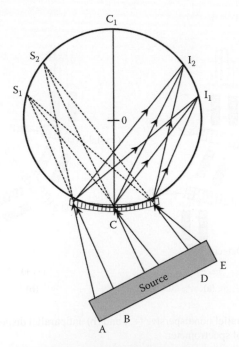

**FIGURE 2.17** Laue spectrometer in line focus (Cauchois) geometry.

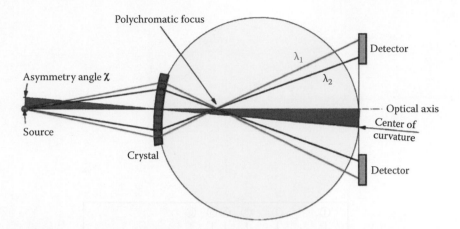

**FIGURE 2.18** Schematic drawing showing the principle of FOCAL geometry. (From Beyer, H.F. et al., 2009, *Spectrochim. Acta B* 64, 736.[111] Copyright 2009. With permission from Elsevier.)

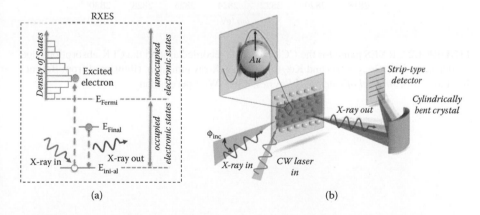

**FIGURE 3.1** (a) Schematic representation of the resonant X-ray emission process, (b) schematics of an experimental setup for resonant X-ray emission spectroscopy employing laser excitation as the external trigger for plasmon determination. (Reprinted from Sá, J. et al., 2013, *Energy Environ. Sci,.* 6, 3584–3588.[33] With permission.)

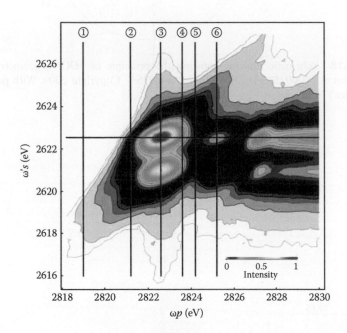

**FIGURE 3.2** RXES plane for the CCl$_4$ molecule recorded around the Cl K-absorption edge by detecting K$\alpha_1$ ($2p_{3/2} \rightarrow 1s$) and K$\alpha_2$ ($2p_{1/2} \rightarrow 1s$) X-ray emission. (Reprinted from Bohinc, R. et al., 2013, *Journal of Chemical Physics,* 139.[35] With permission.)

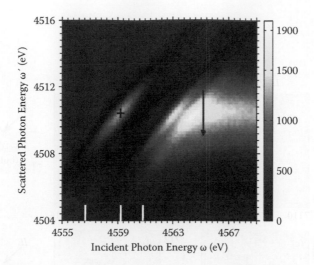

**FIGURE 3.4** RXES plane of $CH_3I$ measured around the $L_3$-absorption edge of I. (Reprinted from Marchenko, T. et al., 2011, *J. Chem. Phys.* 134, 144308.[11] With permission.)

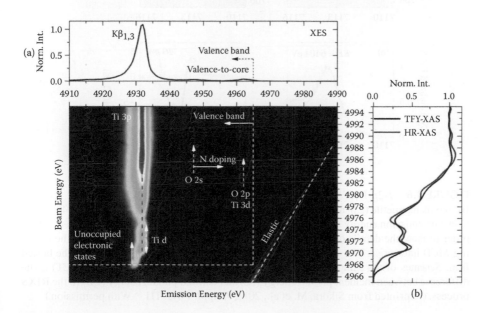

**FIGURE 3.5** $TiO_2$ anatase RXES plane. (a) Nonresonant XES spectrum; (b) TFY- XAS versus HR-XAS extracted at constant emission energy (4931.7 eV). (Reprinted from Szlachetko, J. and Sá, J. 2013, *Cryst. Eng. Comm.* 15, 2583.[56] With permission.)

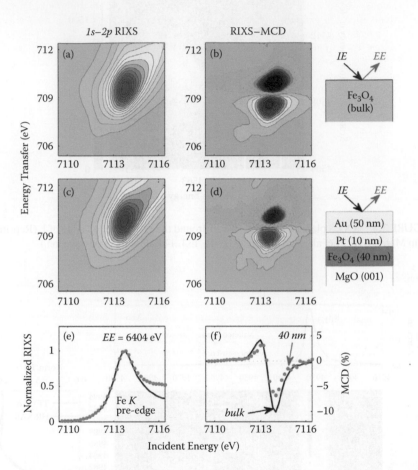

**FIGURE 3.6** *1s2p* RXES plane (a,c) and its magnetic circular dichroism (b,d) acquired from bulk magnetite (a,b) and deeply buried 40 nm thick film (c,d) performed at H1/43.5 kOe. Constant emission energy (diagonal line plots along the RXES planes) are compared (e,f) in order to show the details of normalization (to the maximum of the pre-edge) and the resulting MCD intensities. The solid line represents the bulk sample; circles represent the buried film. Schemes of the sample structures are shown on the right. Energy transfer (ET) is the difference between incident photon energy (IE) and emitted photon energy (EE) in the RIXS process. (Reprinted from Sikora, M. et al., 2012, *J. Appl. Phys.* 111.[58] With permission.)

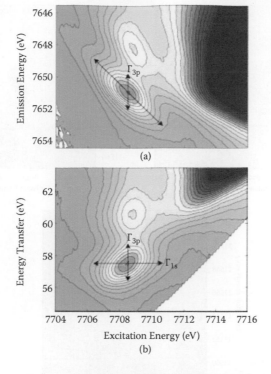

(a)

(b)

**FIGURE 3.7** *1s-3p* RIXS plane of Co(II)-oxide. Also shown are the directions of the core hole lifetime broadenings $\Gamma_{1s}$ and $\Gamma_{3p}$. (a) Emission against excitation energy is shown. (b) Energy transfer against excitation energy is shown. (Reprinted from Kuehn, T.-J., 2011, *Applied Organometallic Chemistry* 25, 577–584.[60] With permission.)

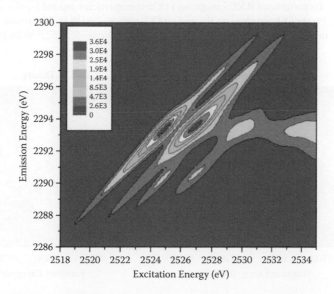

**FIGURE 3.8** The *2p3d* RXES plane for the tetrahedral coordinated Na2MoO4 system around the Mo L$_3$-edge. (Reprinted from Kavčič, M. et al., 2012, *Rev. Sci. Instrum.* 83, 033113.[32] With permission.)

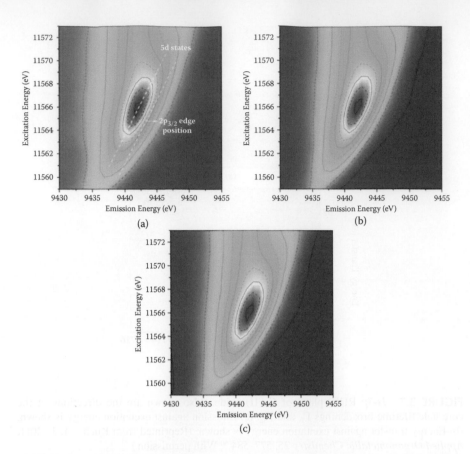

**FIGURE 3.9** Experimental RXES maps on Pt/Co nanoparticles around $L_3$-absorption edge of Pt. (a) Bare Pt; (b) CO adsorbed on Pt; and (c) CO adsorbed on Pt in the presence of a magnetic field (50 mT). (Reprinted from Sá, J. et al., 2013, *Nanoscale* 5, 8462.[74] With permission.)

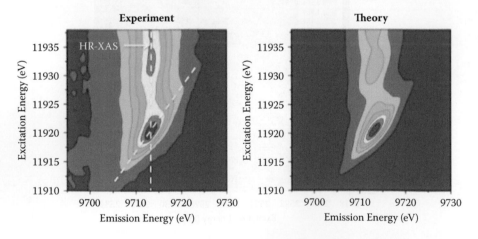

**FIGURE 3.10** RXES map of IRMOF-3-Si-H recorded around the $L_3$-absorption edge of Au. The RXES map is compared with the calculated spectrum on the basis of Slater-type orbitals. (Reprinted from Sá, J. et al., 2013, *RSC Adv.* 3, 12043.[75] With permission.)

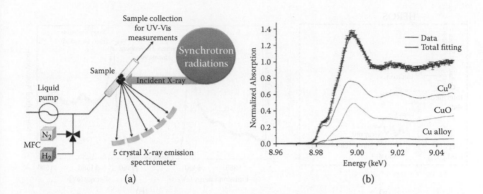

**FIGURE 3.11** (a) Schematic representation of a setup used for *in situ* HR-XAS experiments, (b) linear combination fitting of the normalized HR-XAS spectrum of Pt–Cu/Al$_2$O$_3$ after exposure to 6 mL min$^{-1}$ H$_2$ in the presence of a solution containing 100 ppm of NO$_3$– after 75 min on stream. (Reprinted from Sá, J. et al., 2012, *Catal. Sci. Technol.* 2, 794.[90] With permission.)

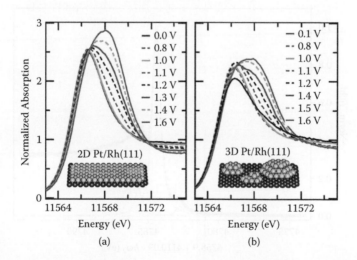

**FIGURE 3.12** *In situ* Pt L$_3$ HR-XAS for 1 ML Pt/Rh(111) in 0.01 M HClO4: (a) 2D Pt film, (b) 3D Pt islands. Spectra were recorded in the order of increasing electrochemical potentials. (Reprinted from Friebel, D. et al., 2012, *J. Am. Chem. Soc.* 134, 9664–9671.[91] With permission.)

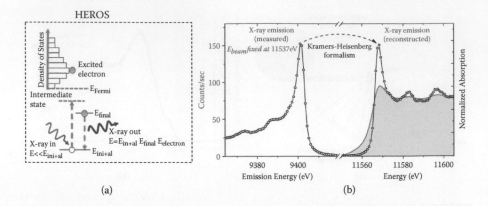

(a)  (b)

**FIGURE 3.13** (a) Schematic representation and energy level diagram for off-resonant scattering process, and (b) the HEROS-XES for the $3d_{5/2}$-$2p_{3/2}$ transition of a Pt foil recorded for a fixed beam excitation energy of 11537 eV. (b) (Blue circles) The reconstructed HEROS-XAS derived from the corresponding HEROS-XES spectrum using the Kramers–Heisenberg formalism. For comparison, the conventional total fluorescence yield spectrum is shown (filled orange area). (Reprinted from Szlachetko, J. et al., 2012, *Chemical Communications* 48, 10898–10900.[100] With permission.)

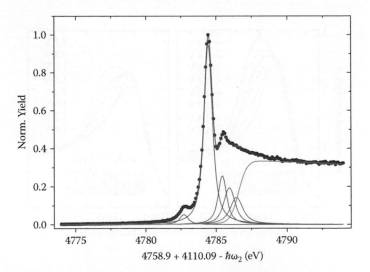

**FIGURE 3.14** The HEROS $L_3$-absorption spectrum of Xe extracted from the measured $2p_{3/2}$-$3d_{5/2}$ RIXS spectrum, which is recorded at 4758.9 eV excitation energy. The spectrum is fitted with a sum of Voigt functions describing the $[2p_{3/2}]$nd and ns discrete excitations followed by a step function corresponding to the $[2p_{3/2}]$-absorption edge. (Reprinted from Kavčič, M. et al., 2013, *Phys. Rev. B* 87, 075106.[101] With permission.)

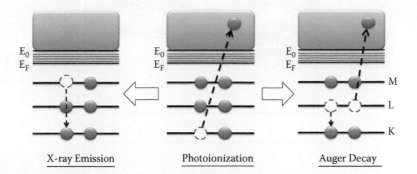

| | | |
|---|---|---|
| $E_0$ | $E_0$ | $E_0$ |
| $E_F$ | $E_F$ | $E_F$ |

X-ray Emission  Photoionization  Auger Decay

**FIGURE 4.1** Schematic energy level diagrams of atomic excitation and relaxation processes. For clarity only the lowest three levels are shown.

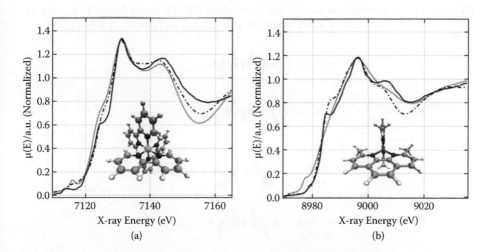

**FIGURE 4.3** (a) The experimental (solid black) and calculated (Muffin tin = dashed, Full potential = dots and dashes) spectrum of $[Fe(bpy)_3]^{2+}$ in water. (b) The experimental (solid black) and calculated (Muffin tin = dashed, Full potential = dots and dashes) spectrum of $[Cu(dmp)_2]^+$ in acetonitrile. The structures of the two complexes are shown inset.

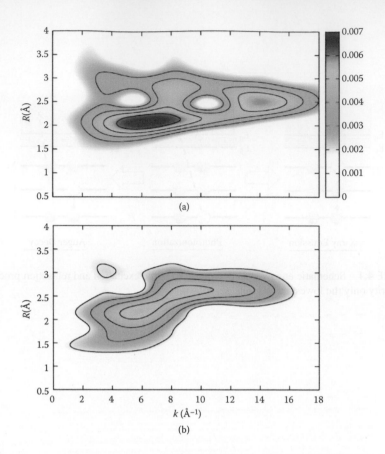

**FIGURE 4.7** A wavelet transform of the $k^2$-weighted EXAFS signal for the Pt $L_3$-edge of the alloy (a) and core-shell (b) Ru-Pt bimetallic nanocatalysts supported on $\gamma$-$Al_2O_3$.[108,109]

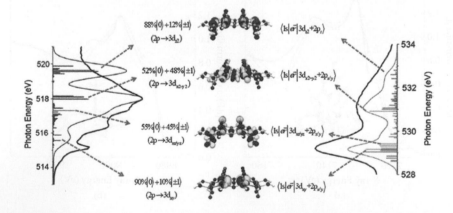

**FIGURE 4.10** Calculated B3LYP/ROCIS V L-edge (red) and O K-edge (green) spectra of the $V_{10}O_{31}H_{12}$ cluster model. The dominant states are assigned in terms of magnetic sublevels, as well as one-electron excitations contributions. The local pentacoordinated environment of vanadium centers is illustrated in the corresponding MO pictures. (Reproduced from Maganas, D. et al., 2013, *Phys. Chem. Chem. Phys.* 15, 7260–7276.[146] With permission from the PCCP Owner Societies.)

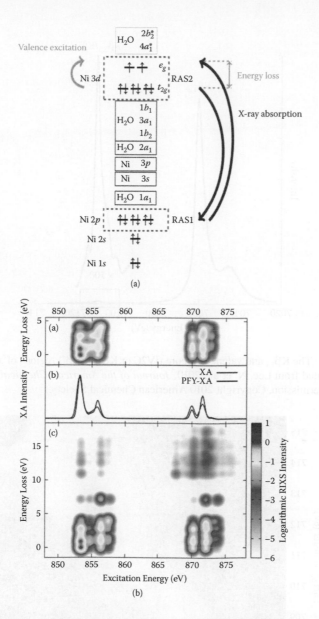

**FIGURE 4.11** Top: Molecular orbital diagram of $Ni^{2+}(H_2O)_6$. For core excitations the active space involves Ni $2p$ and Ni $3d$, partitioned into two subspaces. RAS1 contains the Ni $2p$-orbitals with at most one hole, whereas RAS2 allows all possible electron permutations in the Ni $3d$-orbitals. Ligand-to-metal charge-transfer (LMCT) excitations can be accounted for by introducing ligand orbitals in RAS1 or RAS2. Bottom: (a) The RIXS spectrum corresponding to the RASPT2 X-ray absorption spectrum. (b) The PFY-XA spectrum derived by integrating the RIXS spectrum over all calculated fluorescence decay channels for each excitation energy. (c) RIXS spectrum including charge-transfer (CT) excitations by extending the active space to ligand orbitals. (Figures reproduced from Josefsson, I. et al., 2012, *J. Phys. Chem. Lett.* 3, 3565–3570.[163] With permission.)

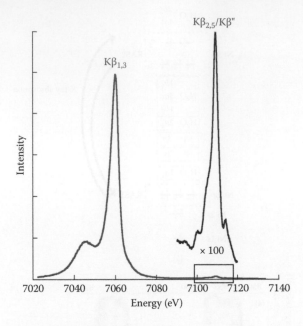

**FIGURE 5.1** The $K\beta_{1,3}$ and valence-to-core (iV2C or $K\beta_{2,5}$ + $K\beta''$) regions of an XES spectrum. (Reprinted from Lee, N. et al., 2010, *Journal of the American Chemical Society* 132, 9715.[9] With permission. Copyright 2010 American Chemical Society.)

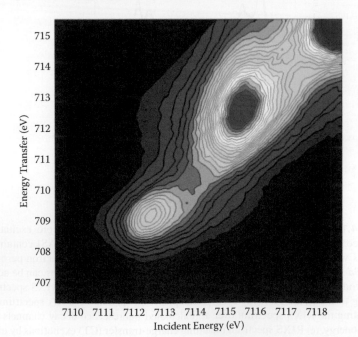

**FIGURE 5.2** A representative *1s2p* RXES plane. The *x*-axis corresponds to the incident energy and the *y*-axis to the energy transfer process, which yields an L-edge-like final state.

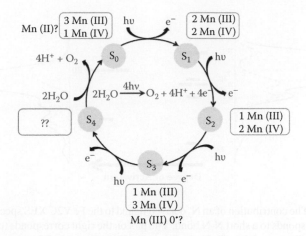

**FIGURE 5.3** The S-State (or Kok cycle) in the oxygen-evolving complex of Photosystem II.

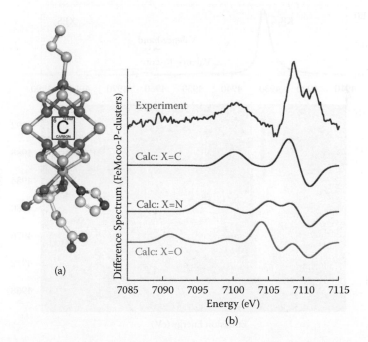

**FIGURE 5.5** (a) The structure of the FeMoco cluster. (b) The experimental V2C XES difference spectra MoFe protein minus the p-cluster contribution. The calculations with a central C, N, or O are shown below. (Adapted from Lancaster, K.M. et al., 2011, *Science* 334, 974.[22] Copyright 2011 American Association for the Advancement of Science.)

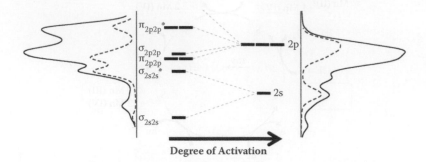

**Degree of Activation**

**FIGURE 5.6** The contribution of an $N_2$ derived ligand to the Fe V2C XES spectrum. The plot on the left corresponds to a short N-N bond. The plot on the right corresponds to a cleaved bis-nitride species. (Reprinted from Pollock, C.J. et al., 2013, *Journal of the American Chemical Society* 135, 11803.[31] With permission. Copyright 2013 American Chemical Society.)

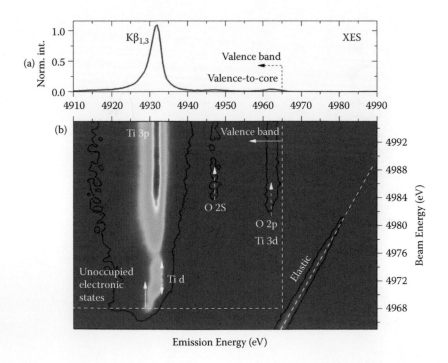

**FIGURE 6.1** TiO$_2$ anatase RIXS map (b) and nonresonant XES spectrum (a). (Adapted from Szlachetko, J. and J. Sá, 2013, *Cryst. Eng. Comm.*, 15, 2583–2587.[5] With permission.)

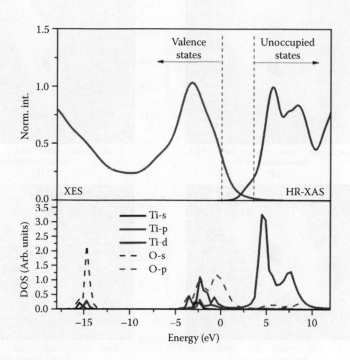

**FIGURE 6.2** Electronic structure of $TiO_2$ anatase extracted from experimental RIXS map (top) and computed DOS (bottom). (Adapted from Szlachetko, J. and J. Sá, 2013, *Cryst. Eng. Comm.*, 15, 2583–2587.[5] With permission.)

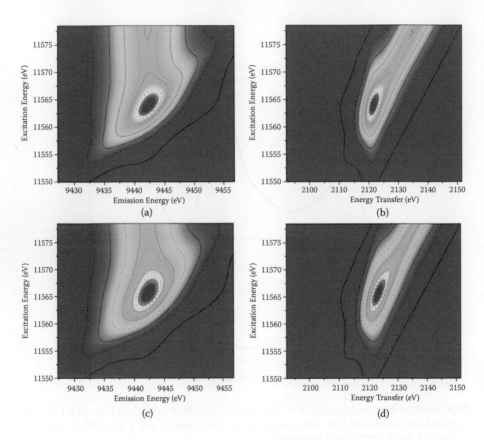

**FIGURE 6.4** Calculated RIXS maps of Pt $L_3$-edge. (a) RIXS of bare $Pt_6$; (b) energy transfer RIXS of bare $Pt_6$; (c) RIXS of CO adsorbed atop on $Pt_6$; (d) energy transfer RIXS of CO adsorbed atop on $Pt_6$.

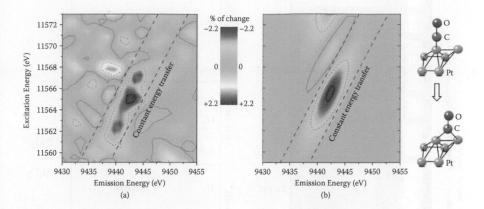

**FIGURE 6.5** Pt L$_3$-edge Δ-RIXS due to the presence of a 50 mT magnetic field on Pt on Co with adsorbed CO (Field OFF–Field ON). (a) Experimental map differences measured *in situ*; (b) calculated map differences. (Reproduced from Sá et al., 2013, *Nanoscale* 5, 8462.[40] With permission.)

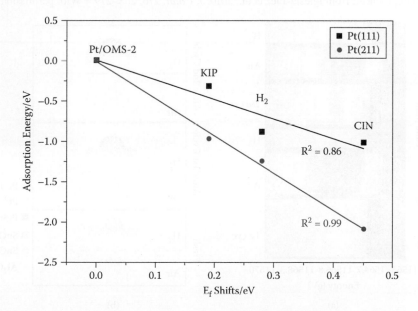

**FIGURE 6.6** Correlation between the experimentally measured Fermi level shifts and the DFT calculated adsorption energies for Pt(111) (black) and Pt(211) (red) surfaces. (From Manyar et al., 2013, *Catal. Sci. Technol.* 3, 1497–1500.[50] With permission.)

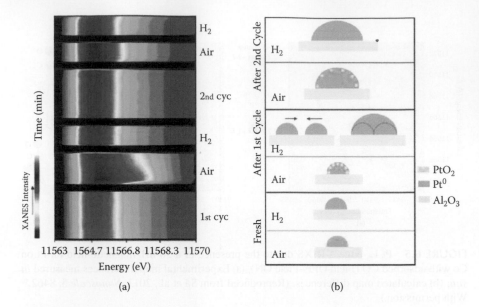

(a)             (b)

**FIGURE 6.7** (a) Intensity contour map of the HERFD-XANES of the Pt/Al$_2$O$_3$ catalyst acquired at 600°C during the first two propane dehydrogenation-regeneration cycles. (b) Schematic model of Pt species during the different treatment stages for the Pt/Al$_2$O$_3$ catalyst. (Reproduced from Iglesia-Juez et al., 2010, *J. Catal.* 276, 268–279.[56] With permission.)

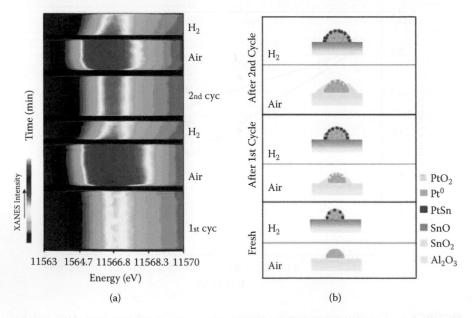

(a)             (b)

**FIGURE 6.8** (a) Intensity contour map of the HERFD-XANES of the Pt-Sn/Al$_2$O$_3$ catalyst acquired at 600°C during the first two propane dehydrogenation-regeneration cycles. (b) Schematic model of Pt species during the different treatment stages for the Pt-Sn/Al$_2$O$_3$ catalyst. (Reproduced from Iglesia-Juez et al., 2010, *J. Catal.* 276, 268–279.[56] With permission.)

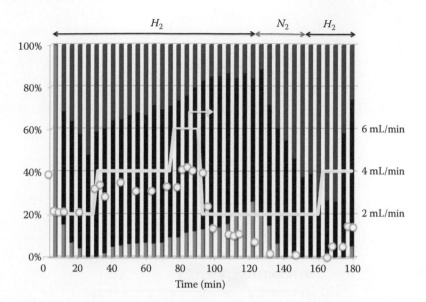

**FIGURE 6.9** Deconvolved Cu K-edge HERFD-XAS spectra of Pt-Cu/Al₂O₃ and nitrate conversion versus gas flow as a function of time stream. (Black) $Cu^0$, (red) CuO, (blue) Pt-Cu alloy; (yellow) gas flow; (′) nitrate conversion. (Adapted from Sá, J. et al., 2012, *Catal. Sci. Technol.* 2, 794–799.[69] With permission.)

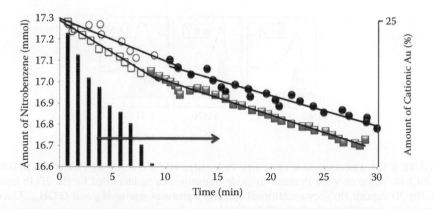

**FIGURE 6.10** Nitrobenzene hydrogenation over Au/CeO₂ catalysts with different concentrations of cationic gold ($Au^{3+}$). Bar plot represents $Au^{3+}$ fraction determined by Au L₃-edge HERFD-XANES, and points represent nitrobenzene conversion. (Red) Au/CeO₂ pretreated at 60°C, and (black) Au/CeO₂ pretreated at 100°C. Reaction carried out under 10 bar H₂, 100°C, and with a stirring rate of 1500 rpm.

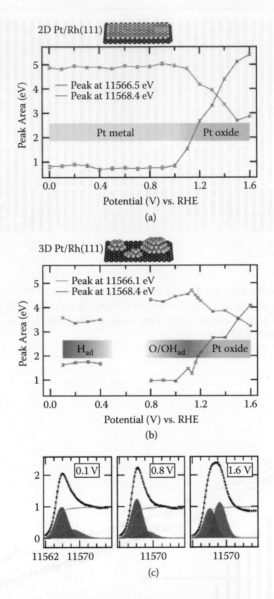

**FIGURE 6.12** Deconvolution of *in situ* Pt L₃-edge HERFD-XAS for Pt/Rh(111) in 0.01 M HClO4. While only the Pt metal-to-oxide transition can be identified for the 2D Pt layer (a), the 3D deposit (b) shows additional spectral signatures due to H$_{ad}$ and O/OH$_{ad}$. Three representative fitting results from 3D Pt/Rh(111) are shown in (c) using the same colors as in (b) for the peak areas, and a gray line for the arctangent function. (Reprinted from Friebel et al., 2012, *J. Am. Chem. Soc.* 134, 9664–9671.[74] With permission. Copyright 2013, American Chemical Society.)

# 4 Theoretical Approaches

## Thomas J. Penfold

## CONTENTS

## 4.1 INTRODUCTION

Core hole spectroscopy is a powerful tool for interpreting the electronic, magnetic, and geometric structure of molecules, solutions, proteins, or solids. The complex mechanism behind the origin of these spectra means that their analysis invariably requires detailed theoretical simulations, which have traditionally been performed using multiple scattering[1-3] or multiplet theories.[4-8] However, rapid experimental developments, as well as continuous improvements in the instrumentation are enhancing the sensitivity of core hole experiments and consequently finer details of the spectra are being uncovered.[4,9,10] In addition, time-resolved experiments with femtosecond temporal resolution[11-14] and picosecond second-order spectroscopies[15] can now be achieved. Finally, the advances brought about by the X-ray free electron lasers [16] offer exciting possibilities for nonlinear core level spectroscopies[17] and for achieving a temporal resolution of ≈10 fs, making it possible to unravel detailed insights into fundamental processes as they evolve in real time. Importantly, these developments call for high-level theoretical approaches so that one may accurately interpret the experimental results.

The theory of core level spectroscopies has advanced significantly in the last decade, and although there exists little disagreement about the physical processes that need to be incorporated, the major challenge is making sufficiently accurate approaches efficient enough (i.e., in terms of computational expense), that they can be routinely used to calculate a wide variety of spectra. This chapter is intended to give an introduction into the state-of-the-art theory of core level spectroscopies, the different tools available for interpreting spectra including their advantages and limitations. Toward this goal, the chapter is organized in the following manner. First, the basics of X-ray absorption and scattering are introduced, proceeded by data reduction schemes critical for the comparison between simulations and experiment. Following this, approaches for obtaining qualitative interpretations of the spectra are described, before the various methodologies for explicitly calculating the spectra, including; multiple scattering (MS) theory, full-potential approaches, multiplet theory, time-dependent density functional theory (TDDFT), and post-Hartree–Fock approaches, are outlined. This chapter is by no means an exhaustive description for the theory of core level spectroscopies and for further reading, a number of books and review articles are highly recommended.[1,7,9,18–25]

## 4.2   BASICS OF X-RAY ABSORPTION AND SCATTERING

An experiment measures the transition rate between an initial and some final state of the system, which may be expressed using second-order perturbation theory as:

$$
F(E) \sim \frac{2\pi}{\hbar} \sum_f \left| \left\langle f \left| \hat{H}_{int} \right| i \right\rangle + \sum_n \frac{\left\langle f \left| \hat{H}_{int} \right| n \right\rangle \left\langle n \left| \hat{H}_{int} \right| i \right\rangle}{E_i - E_n - i \frac{\Gamma_n}{2}} \right|^2
\tag{4.1}
$$

Here, the first-order resonant terms involve absorption and correspond to the matrix elements of X-ray absorption spectra (XAS), while first-order excitations that are nonresonant give rise to scattering. Second-order terms involve the decay of an intermediate state ($n$) following excitation of the initial, usually ground, state and corresponds to the matrix elements for X-ray emission (XES), resonant inelastic X-ray scattering (RIXS), and Auger spectroscopy (Figure 4.1).

The matrix elements in Equation (4.1) depend on the interaction Hamiltonian ($\hat{H}_{int}$) expressed:

$$
\hat{H}_{int} = \sum_{i=1}^{N} \left[ \frac{e}{m} A(r_i) \cdot p_i + \frac{e^2}{2m} A^2(r_i) + \frac{e\hbar}{2m} s_i \cdot \nabla \times A(r_i) - \frac{e^2 \hbar}{(2mc)^2} s_i \right.
$$
$$
\left. \cdot \frac{\partial A(r_i)}{\partial t} \times A(r_i) \right]
\tag{4.2}
$$

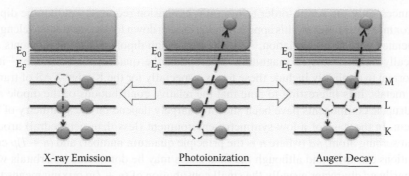

$E_0$        $E_0$        $E_0$

$E_F$        $E_F$        $E_F$

M

L

K

X-ray Emission      Photoionization      Auger Decay

**FIGURE 4.1** **(See Color Insert.)** Schematic energy level diagrams of atomic excitation and relaxation processes. For clarity only the lowest three levels are shown.

where $p_i$ is the momentum of the $i^{th}$ electron, $A$ is the vector potential of the electromagnetic field, and $s_i$ are the Pauli matrices. This can be collected into four principle components:

$$\hat{H}_{int} = \hat{H}_1 + \hat{H}_2 + \hat{H}_3 + \hat{H}_4 \qquad (4.3)$$

where

$$\hat{H}_1 = \sum_{i=1}^{N} \frac{e}{m} A(r_i) \cdot p_i \qquad (4.4)$$

describes absorption and emission processes, and

$$\hat{H}_2 = \sum_{i=1}^{N} \frac{e^2}{2m} A^2(r_i) \qquad (4.5)$$

describes nonresonant scattering. For resonant transitions, $A^2(r_i)$ is much smaller than $A(r_i) \cdot p_i$ and is usually neglected. $\hat{H}_3$ and $\hat{H}_4$ involve the spin matrices and are responsible for nonresonant magnetic scattering[26] and are not considered further within this chapter.

The $A(r_i) \cdot p_i$ term may be rewritten as a wave vector in the form $A_0(e^{i(k \cdot r - \omega t)} + e^{-i(k \cdot r - \omega t)})\varepsilon \cdot p$ where $\varepsilon$ is the polarization vector. The exponential function, $e^{ik \cdot r}$ is then expanded as a Taylor series:

$$e^{ik \cdot r} = 1 + ik \cdot r - \frac{1}{2}(k \cdot r)^2 + \cdots \qquad (4.6)$$

The terms on the right-hand side represent the electric dipole (zeroth order), magnetic dipole and electric quadrupole (first order), magnetic quadrupole and electric octupole (second order), respectively. Within the long wavelength approximation, valid when the wavelength of the radiation is much larger than the interatomic distances, only the zeroth-order term in this expansion required yielding the dipole approximation. However, this approximation breaks down for the shorter wavelengths associated with X-ray radiation,[27,28] and even though dipole transition moments are typically three orders of magnitude larger than the quadrupole transitions,[7] it is important to explicitly include these terms, especially for the K-edge XAS of transition metals. It is interesting to note that the relative contributions of the dipole and quadrupole components have been shown to largely depend on the symmetry of the system. In the case of a low-symmetry environment (less than octahedral) around the absorbing atom, $nd$ (where $n$ is the principle quantum number) and $(n + 1)p$ contributions can mix, and although the final states may be dominated by orbitals with primarily $nd$ character, usually the small contribution of $(n + 1)p$ mixing means that the dipole component dominates the transition matrix element.[29]

For experiments that detect fluorescence as a function of incident energy (e.g., XES and RIXS), the transition matrix elements should be calculated using second-order perturbation theory, that is, the Kramers–Heisenberg equation expressed:

$$F(E) \sim \frac{2\pi}{\hbar} \sum_f \left| \sum_n \frac{\langle f|\hat{H}_1|n\rangle \langle n|\hat{H}_1|i\rangle}{E_i - E_n - i\frac{\Gamma_n}{2}} \right|^2 \qquad (4.7)$$

This also describes the processes involved in high-resolution X-ray absorption spectroscopy, for which the XAS spectrum is recorded in fluorescence yield taking advantage of a particular X-ray emission decay channel.[30,31] These experiments have the advantage that the spectral lifetime broadening is associated with the final state, which is significantly smaller than the $1s$ core hole associated with traditional measurements, yielding a higher energy resolution.[4] In such measurements the spectral line shape is dominated by direct transitions, however, interference effects among the intermediate states which decay into the same final state can affect the spectral weight.[32,33] If the matrix elements of the interference terms are small, which is usually the case for hard X-rays, Equation (4.8) can be rewritten as[34]:

$$F(E) \sim \frac{2\pi}{\hbar} \sum_f \sum_n \left| \frac{\langle f|\hat{H}_1|n\rangle \langle n|\hat{H}_1|i\rangle}{E_i - E_n - i\frac{\Gamma_n}{2}} \right|^2 \qquad (4.8)$$

This modification neglects interference terms and only takes into account the direct terms of the second-order process,[34] in essence giving the emission weighted by the absorption matrix elements.[35]

## 4.3   DATA REDUCTION

The accurate reduction of experimental data is a critical step for comparison to simulations and a thorough understanding of the theory is a prerequisite for a meaningful analysis. During a transmission detection experiment, one records the absorption coefficient, $\mu(E)$, however, theoretically $\chi(E)$ is usually calculated. The relationship between the two is expressed:

$$\chi(E) = \frac{\mu(E) - \mu_0(E)}{\mu_0(E)} \qquad (4.9)$$

where $\mu_0(E)$ is the atomic background. This following section briefly outlines the data reduction methodology for traditional (transmission) XAS measurements before describing the data treatment for fluorescence yield measurements.

### 4.3.1 XANES AND EXAFS

In order to reduce $\mu(E)$ into $\chi(E)$, for traditional transmission XAS data, there are three key steps: (1) pre-edge background removal, (2) edge-jump normalization, and (3) background subtraction. These steps have been discussed at length elsewhere and therefore only the key points are addressed here. For a more detailed description readers are referred to previous reviews[36,37] and methods.[38,39]

Regarding the edge-jump normalization, it is important to recognize that the relationship between $\mu(E)$ and $\chi(E)$ given in Equation (4.9) depends on the atomic background, $\mu_0(E)$, which in general cannot be either measured or calculated accurately. Consequently, normalization is usually performed by dividing by an energy-independent value $\Delta\mu_0(E_0)$, that is, the edge jump. Using this, Equation (4.9) becomes

$$\chi(E) = \frac{\mu(E) - \mu_0(E)}{\Delta\mu_0(E_0)} \qquad (4.10)$$

$\Delta\mu_0(E_0)$ is obtained by fitting a pre- and post-edge polynomial, which are extrapolated to $E_0$. $\Delta\mu_0(E_0)$ is then the difference between the pre- and post-edge functions at this energy, as shown in Figure 4.2a. Although $E_0$, the ionization potential, is at this stage arbitrary, it does not have a large bearing on the results, as long as it lies reasonably close to the real ionization potential of the system. However, it is important to note that because $\mu_0(E)$ is slightly energy dependent, normalizing by $\Delta\mu_0(E_0)$ introduces an attenuation into $\chi(E)$ which is roughly linear in energy. This is usually accounted for by a small (typically about 10%) correction to the Debye–Waller factor, sometimes called the *McMaster correction*.[41]

The final step in isolating the fine-structure oscillations, background removal, is the most important and in contrast to the normalization step, the energy dependence of $\mu_0(E)$ is required. While various ways to approximate the background function exist, the most commonly used approach is the iterative background removal scheme (*autobk*[42]). This takes advantage of the fact that $\mu_0(E)$ contains principally low-frequency components. Using a Fourier transform, a piecewise spline is iteratively fitted to the data such that, when the spline is subtracted, the residual has its low-frequency Fourier components minimized below a certain cutoff distance, usually the first coordination shell distance.

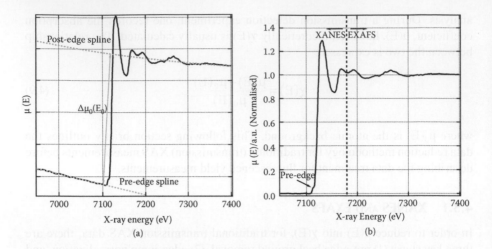

**FIGURE 4.2**    (a) The Fe K-edge absorption edge of nitrosylmyoglobin[40] shown with the pre- and post-edge splines and the edge jump $\Delta\mu_0(E_0)$. (b) The normalized spectrum with the important regions (pre-edge, XANES, and EXAFS) highlighted.

### 4.3.2    Fluorescence Yield Measurements, XES and RIXS

In contrast to traditional data acquisition schemes, it is sometimes advantageous to measure $\mu(E)$ by monitoring processes that are proportional to the absorption coefficient, such as the X-ray total fluorescence yield (TFY) or the total (Auger) electron signal emitted by the sample.[43,44] This is particularly useful when the signal of interest contributes only a small fraction to the total absorption, or when the sample transmission is very large. The fluorescence and Auger signals are due to the absorbing atom only, and if energy-resolving detectors are used, the elastically scattered photons can be discriminated against, resulting in a close to background-free measurement. However, it is important to recognize that recording the X-ray absorption spectrum in this manner is equivalent to performing a second-order measurement, and therefore not necessarily exactly equivalent to traditional XAS.

Normalizing measurements involving fluorescence detection is achieved by dividing by the incoming X-ray intensity $I_0$. However, one must also pay careful attention to attenuation of the signal by self-absorption, arising from the reabsorption of an escaping X-ray and from saturation effects, that is, when $\mu_{FY}$ is much larger than $\mu_{background}$.[45] Corrections for self-absorption have been implemented for EXAFS[46] and RIXS spectra.[47,48]

## 4.4    A QUALITATIVE INTERPRETATION

Often the first step in interpreting core level spectra can be identifying changes and/or trends which occur in a key spectral region, such as the position of the edge (i.e., ionization potential), the pre-edge, the whiteline intensities, and/or the EXAFS oscillations. Changes in these regions can give important insight into the nature of the system under study without the need for calculations. While such analysis should, in the absence of theory, be treated with care these approaches can be particularly useful when used to represent trends as a function of some variable, for example, temperature in catalytic samples.

### 4.4.1 PRE-EDGE AND ELECTRONIC TRANSITIONS

Qualitative interpretations of the valence electronic structure of the system may be obtained using the integrated spectral weight of the spectrum.[49] For example, within the dipole approximation the integrated intensity of the Fe $L_{2/3}$-edge intensity is proportional to the Fe d-character in the valence orbitals on the metal.[50] This interpretation is possible by virtue of the sum rules, and relating such quantities to the orbital and spin moment[51–53] have played a major role in X-ray absorption and X-ray magnetic circular dichroism studies.

The derivation of such rules is significantly more complex for second-order spectroscopies,[20] due to possible interference effects between intermediate states that change the spectral weights. However, when interference effects are small, the sum rules will still hold and can be derived,[54,55] since the decay process has effectively been decoupled from the excitation.

### 4.4.2 EDGE SHIFTS

One of the most common approaches for interpreting core hole spectra is by relating the edge position to the formal charge of the absorbing atom. As the rising edge corresponds to the ionization potential, if the formal charge is larger, it will be harder to remove the electron giving rise to a shift of the edge to higher energies. This can be particularly useful for obtaining a basic understanding of transient (i.e., excited–unexcited) spectra in time-resolved experiments. The so-called *shifted difference spectrum* is calculated as the ground state spectrum shifted to account for the oxidation change minus the ground state spectrum.[56] Using this approach it is possible to ascertain to what extent the features in the transient spectrum arise from a redistribution of electron density within the system following the perturbation and therefore obtain an idea of the magnitude and/or importance of the structural changes.

It is important to note that in addition to the role of the oxidation state of the absorbing atom, the edge positions will depend inversely on the bond lengths of the atoms that coordinate to the absorbing atom.[18,57] Since in some cases there is a strong correlation between the formal charge and the average bond length, it is sometimes difficult to distinguish the two effects and therefore why quantitative simulations are always preferable.

### 4.4.3 EXTENDED FINE STRUCTURE

The EXAFS spectrum is a sum of sine waves with amplitudes that depend on the identity and distance of the nearest neighbor atoms. Therefore, in a qualitative manner, rapid oscillations point toward long path lengths, and a complex oscillatory pattern points to a low-symmetry environment around the atom. In addition, owing to the behavior of the backscattering amplitude (discussed in Section 4.6), if the signal

**TABLE 4.1**

**A Nonexhaustive List of Packages for Simulating Core Level Spectra**

| Code | Method | Spectra | Fit | PBC |
|------|--------|---------|-----|-----|
| FEFF[21] | MST | EXAFS, XANES, ELNES, NRIXS, XES, XMCD | | |
| MXAN[60] | MST | EXAFS, XANES | X | |
| FDMNES[61] | MST/FDM | XANES, XMCD, RXS | X | |
| STOBE[62] | TP-DFT | XANES, XES | | |
| ORCA[63] | LR-TDDFT, ROCIS | XANES, XES, RIXS | | |
| ADF[64] | LR-TDDFT, SF-TDDFT | XANES | | |
| QCHEM[65] | LR-TDDFT, CC, CIS(D) | XANES, XES | | |
| NWChem[66] | LR- and Real-time TDDFT | XANES | | |
| Wien2k[67] | DFT | XANES | | X |
| Abinit[68] | DFT | XANES, XES | | X |
| CP2K[69] | DFT, LR-TDDFT | XANES, XES | | X |
| CTM4XAS[6] | Multiplet | XAS, EELS, XPS, XES, and RIXS | | |

*Note:* Dedicated fitting packages include IFEFFit,[39] SixPack,[58] and FitIt.[59] GAPW = Gaussian augmented plane wave, LR-TDDFT = Linear response time-dependent density functional theory, TP-DFT = Transition potential density functional theory, SF-TDDFT = Spin flip time-dependent density functional theory, ROCIS = Restricted open-shell configuration interaction singles, CC = Coupled cluster, PBC = Periodic boundary conditions.

decays rapidly above the edge this indicates the presence of surrounding atoms of low atomic number. In contrast, for larger atoms the amplitude of the fine structure oscillations can increase as a function of energy above the edge.

These observations are able to give insight into the nature of the system, or trends for a range of similar spectra. However, it is important to keep in mind that the interference of the sine waves can be both constructive and destructive, which can distort the qualitative picture described above.

## 4.5   CALCULATING THE SPECTRA

For simulating core level spectra there are three main methodologies, multiple scattering, multiplet theory, and quantum chemistry approaches, each of which have their own merits and disadvantages as outlined in the following sections. At the most fundamental level all of these approaches aim at addressing the two principle challenges: (1) the need to efficiently describe both the compact (core) initial and diffuse (continuum) final states, and (2) the accurate inclusion of the strong electron correlations that give rise to many-body effects. Toward this goal there are a number of packages capable of simulating the core level spectra, and a nonexhaustive list is shown in Table 4.1.

## 4.5.1 XANES

The analysis of the X-ray absorption near-edge structure (XANES) and EXAFS regions of the spectra are usually treated as separate procedures because, owing to the increased complexity of the physics involved, XANES simulations are usually unable to achieve the same accuracy as EXAFS.[60] Although the gap between the two regions is diminishing, the different information content and the validity of various approximations[1,70] within the two spectral ranges still makes their separate analysis a logical step.

### 4.5.1.1 Multiple Scattering Theory

For solving Fermi's golden rule the major bottleneck is the requirement to calculate the final states. In the continuum these diffuse states are a particular challenge and consequently the most commonly adopted approach is multiple scattering (MS) theory. A characteristic of MS theory is that it is not based upon a basis set expansion of the global wavefunction, as is the case for quantum chemistry calculations. Instead, the global solution is expanded in terms of local solutions of the Schrödinger equation at the energy of interest.[71,72] It is important to recognize that although a physically intuitive picture of a photoelectron scattering off nearest neighbor sites emerges from this approach, MS theory is valid both above and below the ionization threshold, and the scattering order simply represents the extent to which the final state is distorted from the atomic symmetry of the absorber.

In MS theory, the Green's function operator is expressed as:

$$G(r,r',E) = \frac{1}{(\hat{H} - E + i\eta)} \tag{4.11}$$

where $\eta$ is a broadening function. This relates to the density of the system through:

$$\text{Im}\, G(r,r',E) = -\frac{1}{\pi}\rho(r,r',E) \tag{4.12}$$

observing that

$$\rho(r,r',E) = \sum |f\rangle\langle f|\delta(E - E_f) \tag{4.13}$$

we are able to write Fermi's golden rule in the form:

$$\mu(E) \sim -\frac{1}{\pi}\text{Im}\langle i|\varepsilon \cdot r G(r,r',E)\varepsilon \cdot r'|i\rangle \tag{4.14}$$

The power of this approach lies in its ability to separate the propagator, $G(r,r'; E)$, into scattering sequences, each sequence involving individual potential cells with free-particle propagation between the events. Rewriting $G(r,r'; E)$ as[1]

$$G(\boldsymbol{r},\boldsymbol{r}';E) = G^c(\boldsymbol{r},\boldsymbol{r}';E) + G^{sc}(\boldsymbol{r},\boldsymbol{r}',E) \qquad (4.15)$$

makes it possible to show the contributions of the absorbing atom ($G^c$) and the surrounding atoms ($G^{sc}$). $G^{sc}$ can then be represented as a path expansion:

$$G^{sc}(\boldsymbol{r},\boldsymbol{r}',E) = \bar{G}T\bar{G} + \bar{G}T\bar{G}T\bar{G} + \cdots \qquad (4.16)$$

or calculated to all orders using matrix inversion:

$$G^{sc}(\boldsymbol{r},\boldsymbol{r}',E) = \left(1 - \bar{G}T\right)^{-1}\bar{G} \qquad (4.17)$$

Here $\bar{G}$ represents the free Green's function damped by the complex self-energy operator, and T is the scattering matrix.[73]

The separation of the propagator into individual scattering sites imposes that calculations are carried out at nonoverlapping potential cells; however, as demonstrated by Williams and Morgan,[74] there is no limitation on this shape, except the fact that these cells are nonoverlapping. Recent work aimed at extending traditional multiple scattering theory beyond the spherical potentials is discussed in Hatada et al.[72] and Natoli et al.,[75] however, in general this approach is based upon muffin-tin potentials. This approximates the atomic potential as spherical in the region of the atoms and flat (constant) between them, in the so-called interstitial region.[2,3] While this tends to work well in the case of compact systems when the interstitial potential is small, for systems poorly described within this approximation, it has been recently shown that the spectra can be improved using overlapping muffin tin.[76] When the overlap is around 10 to 15% the benefit can remain greater than the error, however, this approach is mathematically incorrect and these false improvements can hide structural or electronic information.[61]

One of the advantages of the Green's function formulation is that it can incorporate inelastic losses and other quasi-particle effects without the necessity of explicit calculation of wavefunctions. Here, the potential is separated into local regions around each atom expressed:

$$V_{coul} + \Sigma(E) = \sum_R v_R(r - R) \qquad (4.18)$$

where the bare Coulomb potential ($V_{coul}$) is modified by the self-energy operator, $\Sigma(E)$, which incorporates the many-body effects and inelastic losses and can be thought of as analogous to the exchange-correlation potential of DFT. The Hedin–Lundqvist self-energy[77,78] in the local density approximation is the most commonly used. This complex operator consists of a real part accounting for the energy dependence of the exchange interaction, and consequently introduces an energy shift in the spectral positions. The imaginary part accounts for the inelastic losses of the photoelectron as it propagates and yields the mean free path ($\lambda$):

$$\lambda = \frac{k}{\text{Im}\left[\Sigma(E)\right]} \tag{4.19}$$

In addition to inelastic losses, sometimes called *extrinsic losses*,[1] the spectrum also contains *intrinsic losses*. These denote many-body excitations, which are analogous to a higher-order excitation expansion space in configuration interaction.[75,79] Recently, there has been extensive work on incorporating many-body effects into MS theory-based approaches with great success. However, their description is beyond the scope of this present chapter and readers are instead recommended the following articles.[70,75,80,81]

Despite the present limitations of the description of the potential, MS theory provides for many cases a good description of X-ray spectra and should always be the first calculation performed. In addition, this approach provides a physically intuitive description of the continuum resonances and can be further interpreted using a shell-by-shell analysis.[82] This consists of gradually increasing the number of atoms around the absorbing atom, so that the appearance of the resonances can be attributed to specific scattering paths.[83,84]

### 4.5.1.2 Full Potential Approaches

Although the muffin-tin approximation is sufficient for large regions of the spectrum, close to the edge in the XANES region of the spectrum the excited electron is strongly sensitive to the fine details of the atomic potential meaning that this spherical approximation is no longer suitable. Such problems are most commonly encountered in the case of open structured systems,[85,86] or when the absorbing atom is not fully coordinated meaning, in both cases, that the approximated interstitial region is large.

The most widely used alternative is the finite difference method near-edge structure (FDMNES) approach by Joly.[61,87] For these calculations, the system is decomposed into three regions, (1) the atomic core, (2) the continuum, and (3) the valence region. The finite difference is used to solve the wavefunction in the important valence region, while the continuum (which has a constant potential) is solved using a basis of Neumann and Bessel functions. Finally, the atomic core is described using spherical harmonics, similar to the approach of the MT potential. However, in this case, the spherical region is smaller than that typically used in the MT approach, and is only used to save the computational expense associated with having the dense grid required to accurately describe the potential close to the nucleus.

The advantage of a full potential description is demonstrated in Figure 4.3 showing the XANES spectrum of two transition metal complexes; $[Fe(bpy)_3]^{2+}$ (Figure 4.3a[88,89]) and $[Cu(dmp)_2]^+$ (Figure 4.3b[56,90,91]), calculated with the FDMNES approach using the full potential and the muffin-tin approximation. For the low-spin spectrum of $[Fe(bpy)_3]^{2+}$ there is close agreement between the muffin-tin and full potential calculations and the experiment. This is because the ferrous iron is fully coordinated (octahedral geometry) and as the Fe-N bond length is only 2.0 Å the interstitial region is small and has little effect on the scattering within this complex.

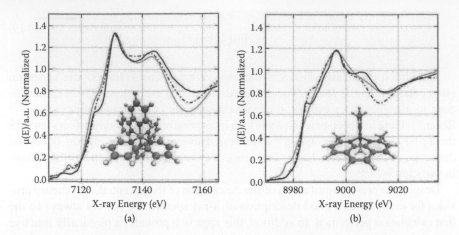

**FIGURE 4.3  (See Color Insert.)** (a) The experimental (solid black) and calculated (Muffin tin = dashed, Full potential = dots and dashes) spectrum of $[Fe(bpy)_3]^{2+}$ in water. (b) The experimental (solid black) and calculated (Muffin tin = dashed, Full potential = dots and dashes) spectrum of $[Cu(dmp)_2]^+$ in acetonitrile. The structures of the two complexes are shown inset.

In contrast, Figure 4.3b shows the XANES spectrum for $[Cu(dmp)_2]^+$. In this case, the absorbing copper atom is in a tetrahedral coordination environment and therefore a larger interstitial region exists. The influence of this approximated interstitial region is highlighted by the poor agreement between the muffin-tin calculation and the experimental spectrum, in comparison to the full potential approach that captures all of the features appearing in the experimental spectrum.

Despite offering a method for calculating the XANES spectrum without restrictions on the form of the potential, the major limitation of the FDMNES approach is that the calculations are computationally expensive. This derives from the maximum angular momentum used to connect the interstitial region to the outer sphere, given by:

$$kr = \sqrt{l_{max}(l_{max} + 1)} \tag{4.20}$$

where $k$ is the photoelectron wave vector and $r$ is the cluster radius. In the absence of symmetry, this limits its application to clusters of not more than 30 atoms.[92]

Alternatively, as shown in Table 4.1, approaches based upon plane waves have also been widely implemented.[68,93–95] The ability of plane waves to describe the diffuse continuum states is well established and the simple one-parameter cutoff to set the accuracy of the basis makes it very appealing. However, a plane wave basis cannot describe the atomic core region of the wavefunction and consequently pseudopotentials have to be used. In this case, the projector augmented wave method is commonly implemented[93] to obtain the transition strengths in the absence of the core electrons. Although the neglect of the core electrons makes the calculations less accurate, it also means that such calculations are less computationally expensive and able to tackle larger problems. However, it is important to stress that it is not easy to transfer pseudopotentials between different computational packages and that they must be tested extensively in different chemical environments. A promising approach for avoiding the use of pseudopotentials are methods based upon a mixed basis set, such as Gaussian augmented plane wave (GAPW).[69,96] The GAPW method uses a Gaussian basis to expand the molecular orbitals and plane waves for the diffuse component of the charge density. This method is suitable for all-electron calculations on several hundreds of atoms and for lattice structures; it can be used in combination with the supercell approach and periodic boundary conditions (PBCs). It has been used to calculate core hole spectra,[97,98] achieving good agreement with the experiment and therefore further investigations in this area could provide an exciting alternative to the methods described above.

## 4.5.2 EXAFS

The extended X-ray absorption fine structure (EXAFS) region includes the high-energy continuum, typically >50 eV above the absorption edge. In this region the oscillatory structure arises from the interference between the outgoing and scattered photoelectron waves. Because the photoelectron wavelength decreases as a function of its kinetic energy, the contributions to the final state wavefunction in the EXAFS region are generally in the vicinity of the absorbing atoms and can therefore be effectively simulated using only the dominant single scattering pathways. In addition, at these excitation energies above the absorbtion edge, the photoelectron is not sensitive to the fine details of the potential and the many body effects can be accounted for in a phenomenological manner within the amplitude reduction factor.[1]

In the context of high-resolution spectroscopy, EXAFS may also be recorded in fluorescence yield,[99] and in this case can be sensitive to different sites of the same absorbing atoms. This can be especially useful in the case of mixed valence systems[100] or for achieving a larger EXAFS range when the spectra of two different absorbing atoms overlap.[101] If the fluorescence arising from the difference spin- or oxidation-state sites are completely separable, the EXAFS spectrum recorded would represent the spectrum of only one particular site, otherwise the spectra is a linear combination of sites at that emission energy and by changing the emission energy one is simply manipulating the coefficients of this combination.[102]

### 4.5.2.1 EXAFS Equation

The EXAFS equation is expressed:

$$\chi(k) = \sum_{\gamma} \frac{N_{\gamma} S_0^2 F_{\gamma}(k)}{k R_{\gamma}^2} e^{-2R_{\gamma}/\lambda(k)} e^{-2\sigma^2 k^2} \sin\left(2kR_{\gamma} + \phi_{\gamma}\right) \quad (4.21)$$

$\gamma$ is the scattering path index with degeneracy $N_{\gamma}$. The half-path distance and the squared Debye–Waller factor are represented by $R_{\gamma}$ and $\sigma^2$, respectively. $\lambda(k)$ is the energy-dependent mean free path and $S_0^2$ is the amplitude reduction factor, which accounts for many-body effects.

Equation (4.21) is comprised of two principle components, an amplitude (the first three terms) and a phase (the final term). The amplitude contains information about the nearest neighbors (coordination), atomic species, and disorder, while the phase component consists of the interatomic distances ($2kR_\gamma$) and a phase shift ($\phi$) expressed as:

$$\phi_\gamma(k) = 2\phi_\gamma^{absorber}(k) + \phi_\gamma^{scatterer}(k) \tag{4.22}$$

This equation describes the changes to the phase associated with the increase in velocity of the photoelectron as it approaches the neighboring nuclei. This is represented pictorially in Figure 4.4. Unfortunately, both the amplitude and the phase depend only weakly on the atomic species and while this resolution may be enhanced by scaling the spectrum in $k$-space, typically by $k$, $k^2$, or $k^3$, this does not generally help to distinguish very similar atoms, such as C and N, meaning that it can be difficult to identify an unknown scatterer with precision from an EXAFS analysis.

This is highlighted in Figure 4.5, showing the backscattering amplitude as a function of the photoelectron energy, $k$. The backscattering has a resonant characteristic and is therefore larger at energies for which the photoelectron equals the orbital energy of a bound electron associated with the backscattering atom. Consequently,

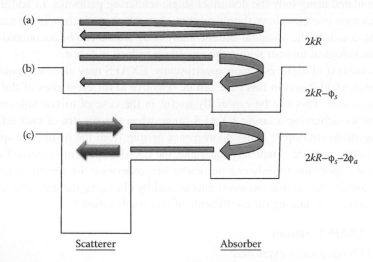

**FIGURE 4.4** Schematic of the contributions to the phase factor in the EXAFS equation. (a) The phase delay is equal to $2kR$, reflecting the time required for the photoelectron to travel out to the scatterer atom and return. (b) The phase decreases by an amount $\phi_s$ because near the scatterer the kinetic energy of the electron is increased thereby decreasing the time required to transverse the $2R$ distance. (c) The phase is decreased by the additional term $2\phi_a$ because near the absorber core the photoelectron kinetic energy is also larger. (Figure redrawn from Köningsberger, D. et al., 2000, *Top Catalysis* 10, 143–155.[37])

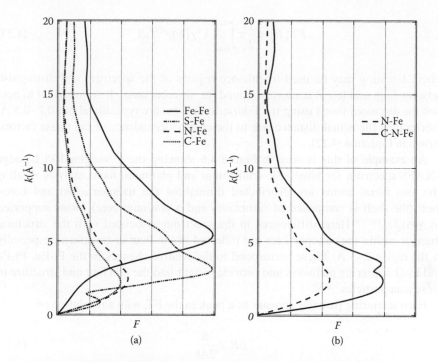

**FIGURE 4.5**   Backscattering amplitude, $F$, as a function of $k$ for single (a) and multiple (b) scattering pathways.

for carbon and nitrogen there are three maxima, one caused by scattering off the $2p$ (3 Å$^{-1}$) orbital electrons, the second by scattering off the $2s$ (4 Å$^{-1}$), and a final one, the weak scattering of the $1s$ electrons (11 Å$^{-1}$). For sulphur, the $3s$- and $3p$-orbitals also contribute (3.5 and 5.5 Å$^{-1}$), while the strong resonance (5 Å$^{-1}$) in the case of iron is the $3d$ electrons. Figure 4.5b shows the enhancement backscattering amplitude along the linear Fe-C-N bond. Here, the structure of the backscattering remains the same but the amplitude is increased owing to the focusing effect.[103]

EXAFS spectra contain information about the average bond length of the surrounding atoms which contribute to the signal; these will be damped by the thermal motions associated to the scattering paths and are encapsulated in the Debye–Waller factor ($\sigma^2$) shown in Equation 4.21. The damping effects of the Debye–Waller factor are most commonly included as a free parameter during the fitting procedure of an EXAFS spectrum (Section 4.6.4), but can be calculated in an *ab initio* manner using the projected density of vibrational states[104,105] or from molecular dynamics simulations.[106]

### 4.5.2.2   Fourier Transform

Usually, the initial step in obtaining a qualitative description of the structure is achieved using a Fourier transform (FT) of the EXAFS signal, which yields a pseudo-radial distribution.[107]

$$FT(R) = \frac{1}{2}\pi \int_{k_{min}}^{k_{max}} k^n \chi(k) e^{i2kR} dk \qquad (4.23)$$

where $k$-scaling may be used to enhance regions of the spectrum and distinguish between high and low Z scatterers around the absorber atom. It is important to note that the distances found using the Fourier transform are typically about 0.2–0.5 Å, shorter than the actual distance due to the energy dependence of the phase factors, shown in Equation (4.22).

An example of this is seen in Figure 4.6 showing the $k^2$ weighted Pt $L_3$-edge EXAFS spectrum for bimetallic (ruthenium and platinum) nanocatalysts in alloy (the two metal atoms are distributed throughout the nanoparticles) and core–shell (the shell is composed of ruthenium and a platinum core) forms supported on $\gamma$-$Al_2O_3$.[108,109] Here, differences in the spectrum associated with the structural changes within the nanoparticles are reflected in the Fourier transform, especially in the region 2–3 Å. These correspond to the relative changes in the Pt-Ru, Pt-Pt, and Pt-O scattering pathways and provide insight into the bonding and structure of these nanoparticles.[108,109]

Each scattering pathway appears as a peak in the FT, with a resolution of:

$$\Delta R \geq \frac{\pi}{2\Delta k} \qquad (4.24)$$

This means, for example, that data collected with a $\Delta k = 11$ Å$^{-1}$ will have a resolution of $\Delta R = 0.14$ Å, and therefore scattering pathways which differ by $\geq \Delta R$ will

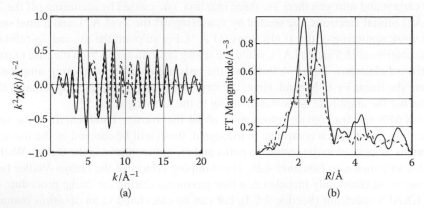

**FIGURE 4.6** (a) Edge-step normalized, background-subtracted EXAFS signal, weighted with $k^2$, for the Pt $L_3$-edge in the alloy and core-shell bimetallic nanocatalysts supported on $\gamma$-$Al_2O_3$. (b) Fourier transform magnitudes of the $k^2$-weighted for the Pt $L_3$-edges in the alloy and core-shell bimetallic nanocatalysts supported on γ-Al2O3. (The data has been reproduced from Frenkel, A.I., 2012, *Chem. Soc. Rev.*, 41, 8163[108]; Alayoglu, S. et al., 2009, *ACS Nano*, 3, 3127–3137.[109] With permission.)

appear as one peak. For systems, which contain many scattering paths of different atomic species, and/or single and multiple scattering pathways that contribute to the same region of $R$-space, an unambiguous assignment of the peaks can be difficult. This is because these overlapping scattering pathways will alter the appearance of the FT and give rise to peaks that derive from a combination of multiple pathways, rather than one specific pathway. This is particularly pertinent if the data is Fourier filtered. Here, certain distances are extracted in $R$-space and then back transformed into $k$-space to simplify the analysis for systems with low symmetry. In such cases, features are assigned and fitted as if they are composed of a single scattering (SS) pathway, when in reality they arise from a mixture of many overlapping single scattering pathways or of MS will give erroneous results.[110]

### 4.5.2.3   Wavelet Transform

The limitation of the FT is that the resolution is only achieved in real space, $R$. However, as shown in Figure 4.5, the backscattering amplitude has a $k$-dependence and therefore it is not only what frequencies (corresponding to the bond lengths) that are present in the EXAFS spectrum, but at what $k$-value they appear that can yield information about the atoms involved in the scattering pathways. To go beyond the one-dimensional approach provided by the FT, one can use a wavelet transform (WT) analysis.[111–113] Here, the infinitely expanded periodic oscillations of the FT are replaced by a local function, a wavelet, which enables one to analyze the components in $k$- and $R$-space, concurrently. This yields a 2D correlation plot in both coordinates (analogous to a time-frequency correlation plot) and will separate the contributions between different scattering pathways at the same distance from the absorbing atom and between the contributions of SS and MS events.

An example of the wavelet transform approach is demonstrated in Figure 4.7 for the EXAFS spectra shown in Figure 4.6a. Here, the rather convoluted Fourier transform (Figure 4.6b), is separated in both the $k$ and $R$ space. The feature at large $k$ can be assigned to the Pt-Pt scattering pathway,[37] while the shoulder at $k = 4$ Å$^{-1}$ and $R = 1.5$ Å in Figure 4.6b can be assigned to the Pt-O pathway. However, when using this as a qualitative analysis, care must be taken because, like Fourier transforms, the wavelet transform is a complex function. Here, we plot the modulus and consequently lose the information about the phase. This means that the two well-separated peaks at $k = 5$ Å$^{-1}$, $R = 2$ Å, and $k = 9$ Å$^{-1}$, $R = 2.5$ Å, respectively, are, in fact, contributions from overlapping Pt-Pt and Pt-Ru pathways, and it is the interference effect between the two that gives the appearance of two bands, more separated than in reality.[110]

### 4.5.2.4   Fitting EXAFS Spectra

The qualitative interpretation provided above using the Wavelet or Fourier transforms cannot provide the full structural interpretation, which is obtained using a fit of the EXAFS spectrum, for which there are numerous packages available (see Table 4.1). Initially, an EXAFS calculation is performed to extract the details of the

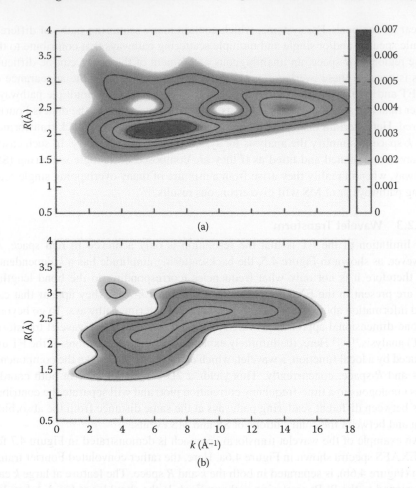

**FIGURE 4.7** **(See Color Insert.)** A wavelet transform of the $k^2$-weighted EXAFS signal for the Pt $L_3$-edge of the alloy (a) and core-shell (b) Ru-Pt bimetallic nanocatalysts supported on $\gamma$-Al$_2$O$_3$.[108,109]

scattering paths. From these the most important are selected and, in addition to variables including the amplitude reduction factor and the Debye–Waller factor, form the parameter space to be optimized. During such procedures it is important to comply with the Nyquist criterion[114]:

$$N_{\text{int}} = \frac{2\Delta k \Delta r}{\pi} + 1 \tag{4.25}$$

describing the minimum ratio of independent points compared to the number of fitting variables for a meaningful fit.

In addition to the structural parameters chosen, $E_0$, the ionization potential is also usually refined during a fit. This is straightforward, although it has to be noted that phase differences between theory and experiment can arise from both errors in $E_0$ or the distance of the scattering pathways. However, since the phase shift decreases, as a function of energy above the edge, a characteristic sign of a relative $E_0$ shift between the calculated and experimental EXAFS signals is a phase discrepancy that *decreases* as $k$ increases. In contrast, a distance error during the fit will be expected to cause a phase difference that *increases* with $k$.

### 4.5.3 ELECTRONIC TRANSITIONS: PRE-EDGE AND $L_{2/3}$-EDGES

In this section we outline the approaches and methodologies for simulating the transitions associated with the valence orbitals below the ionization potential that provide information about the electronic structure of the system. These regions of the spectrum can be simulated within the approaches described in Sections 4.5.1 and 4.5.2; however, the ability to describe these transitions within atom centered (Gaussian) basis sets means that it is possible to extend many traditional electronic structure approaches to core hole excitations. These methodologies have been implemented within several quantum chemistry codes.[63–66]

#### 4.5.3.1 Time-Dependent Density Functional Theory

Time-dependent density functional theory (TDDFT) has seen an explosion in the last decade, owing to the good balance between computational expense and accuracy. This has led to these calculations being extended to core hole excitations and they have been used to study a wide variety of systems.[56,115–120]

TDDFT provides the framework to map the time-dependent Schrödinger equation onto the time-evolving electron density under the perturbation of an external field.[121,122] Although approaches based upon the explicit propagation of the density matrix (real-time TDDFT) under the influence of the time-dependent Fock matrix have been applied to core hole excitations,[123] calculations are generally performed within linear response TDDFT (LR-TDDFT) using Casida's matrix equation written[124,125]:

$$\begin{pmatrix} \mathbf{A} & \mathbf{B} \\ \mathbf{B} & \mathbf{A} \end{pmatrix}\begin{pmatrix} \mathbf{X} \\ \mathbf{Y} \end{pmatrix} = \omega\begin{pmatrix} 1 & 0 \\ 0 & -1 \end{pmatrix}\begin{pmatrix} \mathbf{X} \\ \mathbf{Y} \end{pmatrix} \tag{4.26}$$

where the matrix elements are:

$$A_{ia,jb} = \left(\varepsilon_a - \varepsilon_i\right)\delta_{ai,jb} - C_{HF}\left(ij\middle|r_{12}^{-1}\middle|ab\right) + \left(1 - C_{HF}\right)\left(ia\middle|f_H + f_{xc}\middle|jb\right) \tag{4.27a}$$

$$B_{ia,jb} = -C_{HF}\left(ij\middle|r_{12}^{-1}\middle|ab\right) + \left(1 - C_{HF}\right)\left(ia\middle|f_H + f_{xc}\middle|bj\right) \tag{4.27b}$$

The labels $i$ and $j$ refer to occupied spin orbitals and $a$ and $b$ refer to unoccupied spin orbitals. $\mathbf{X}_{ia}$ and $\mathbf{Y}_{ia}$ represent the particle-hole and hole-particle components. The solution of Equation (4.26) yields the transition frequencies, $\omega$, and transition amplitudes $\mathbf{X}_{ia}$, $\mathbf{Y}_{ia}$, which are used to calculate the dipole and quadrupole moments. Quite commonly, Equation (4.26) is simplified by neglecting the $\mathbf{B}$-matrix (or alternatively constraining $\mathbf{Y}$ to be zero), which yields:

$$AX = \omega X \tag{4.28}$$

the so-called Tamm–Dancoff approximation (TDA).[126] The energetics within the TDA are usually indistinguishable from the full treatment, and decoupling the excitations from the de-excitations simplifies the working equations.

From a practical perspective, TDDFT within the standard approach requires computing all other valence and core excitations before reaching the excitations of interest (usually the $1s$). This is clearly computationally expense, but can be overcome by projecting onto a manifold of single core to valence excitations.[119] Although it neglects the coupling of the core-excited K-edge states to L- and M-edges, as well as valence excited states, it is a good approximation due to the large energy difference between the $1s$ core level and all other orbitals.[27,115,127] Importantly, these calculations yield transitions that give at least a qualitative, and often quantitative, description of the pre-edge region of the spectrum and have been applied to calculate both static and time-resolved XAS. In the case of time-resolved experiments, the transient (excited–unexcited) spectrum requires not only accurate energies, but also oscillator strengths and therefore represents a rigorous challenge for the method. This was recently used to calculate the transient spectrum for a copper(I)-phenanthroline complex (Figure 4.8).[56] Here the transitions in the

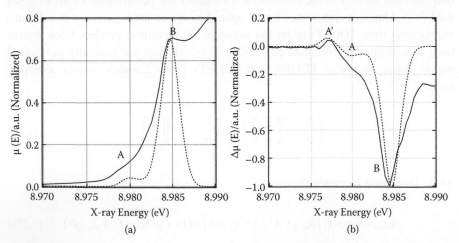

**FIGURE 4.8**    (a) The pre-edge region of the Cu K-edge spectrum of [Cu(dmp)$_2$]$^+$ and the oscillator strengths calculated using TDDFT with a 1.89 eV Lorentzian broadening. (b) The simulated pre-edge transient spectrum (dotted line, singlet-triplet) using TDDFT in comparison to the experimental transient (solid line).[56]

ground state spectrum (A and B) were assigned as metal($1s$)-to-ligand charge transfer (MLCT) and $1s$-$4p$ transitions, respectively. The transient (excited–unexcited) spectrum exhibits a new positive feature corresponding to the hole in the $3d$-orbitals created by photoexcitation of the complex. The negative features, which correspond to A and B in the transient spectrum, arise from the oxidation shift of the transitions owing to the Cu(I) becoming Cu(II).

It is important to recognize that for all TDDFT simulations the exchange-correlation ($x$-$c$) functional, which describes the many-body effects, is crucial for obtaining an accurate description of the excited states. While a wide range of different functionals exist, generally the generalized gradient approximation (GGA) functionals (e.g., BP86,[128,129] PBE[130]) or hybrid (e.g., B3LYP,[131] PBE0[132]) functionals are the most commonly used. When performing these calculations there are two important factors which users should be aware of. First, the asymptotic limits ($r \rightarrow 0$ and $\infty$) of the potential are incorrect for most standard $x$-$c$ functionals.[133] This failure stems from the approximate exchange within the $x$-$c$ functionals and is associated with the self-interaction error (SIE), as the exchange term does not necessarily cancel out the term of the Coulomb, which includes the interaction of the electron with itself as is the case in Hartree–Fock theory.[134] The effect of this for high-lying excitations, such as Rydberg states, is well documented and results in absolute excitation energies that are too low in comparison to experimental.[135] The same is also true for core excitations, and the absolute transition energies must typically be shifted by 50–300 eV to compare with experiment. However, although it is desirable to have the correct absolute excitation energies, it is the relative energies that remain the most important, and these are usually relatively unaffected. In addition, Debeer-George and coworkers[27,115,136] have performed detailed calibrations of these errors, showing that for a particular atomic edge, the energy shift is almost independent of the surrounding ligands within the limit of the same calculation variables (i.e., basis set, functional etc.).

Second, as has been extensively studied for valence excitations, excited states with charge transfer (CT) character can be problematic when simulating core hole excitations within TDDFT.[137,138] This arises from the inherent spatially local nature of GGA functionals,[139] and in such cases, it is important to include a percentage of nonlocal Hartree–Fock exchange (i.e., hybrid functional). This component of nonlocal exchange enters as the second term of Equation (4.27a) and has a Coulombic characteristic, corresponding to the electrostatic attraction between the electron and the hole.

### 4.5.3.2 Multiplet Theory

The one-electron description of core level spectra is often sufficient for understanding the main features in a wide variety of cases, as demonstrated by the success of the approaches described above. However, this approximation can break down and this is most commonly encountered when simulating metal $L_{2,3}$-edges.[140] The reason is the overlap between the core and valence wavefunctions, or more precisely the $2p$-$3d$ two-electron integrals. These strongly modify the empty $3d$ density of states,

giving rise to so-called multiplet effects. Traditionally, these effects have been simulated using approaches based upon atomic multiplet theory.[4,5,7]

The basis of atomic multiplet theory[141] is to address the main problem of accounting for electron correlation by splitting the electron–electron repulsion into spherical
and nonspherical components. The former is then added to the atomic Hamiltonian
to form the average energy of a configuration:

$$\hat{H}_{atom} = \sum_N \frac{p_i^2}{2m} + \sum_N \frac{-Ze}{r_i} + \left\langle \sum_{pairs} \frac{e^2}{r_{ij}} \right\rangle \tag{4.29}$$

where the final term is the spherical component of the electron–electron repulsion.
The nonspherical part is defined:

$$\hat{H}_{ee}^{ns} = \sum_{pairs} \frac{e^2}{r_{ij}} - \left\langle \sum_{pairs} \frac{e^2}{r_{ij}} \right\rangle \tag{4.30}$$

and in conjunction with spin-orbit coupling yields the different terms within
atomic configurations.

For a successful description of core hole spectra, one is usually required to include
the effect of the surrounding ligands (ligand-field multiplet) and charge transfer
(charge-transfer multiplet) effects on top of the atomic Hamiltonian. The ligand field
is simply an electrostatic field on top of the atomic Hamiltonian and charge-transfer
effects involve the coupling of two or more electron configurations into the initial
and/or final states.[5,7]

Multiplet-based approaches have been widely and successfully used to describe
the features appearing in many core hole spectra and their main advantage is their
high computational efficiency.[29,50,142,143] However, the principle drawback is the reli-
ance on a large number of semiempirical parameters that describe the effects of
electron–electron repulsion, the ligand environment, and covalency. In addition,
group theory is heavily used to simplify the two-electron integrals.

### 4.5.3.3    Post Hartree–Fock Methods

Approaches based upon TDDFT represent an excellent balance between accuracy
and computational expense; however, they are usually unsuitable for simulating
spectra in which multiplet effects and spin-orbit coupling become important. First,
the standard formulation of TDDFT, within the adiabatic approximation does not
include spin-flip excitations and second, in order to properly account for multiplet
effects within core level spectra, higher-order excitations and even multideterminental wavefunctions are required.[144] While, as discussed in the previous section, it is

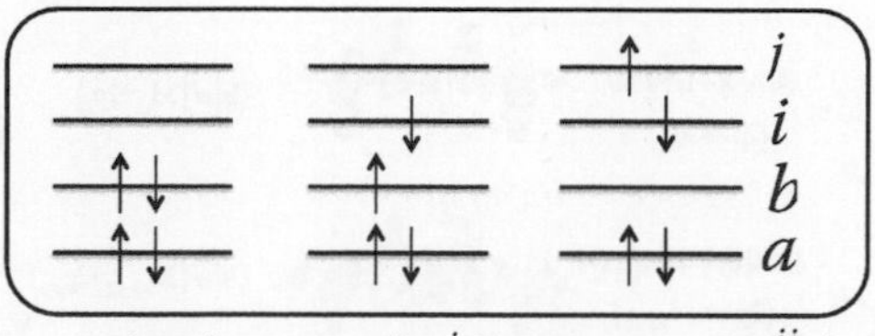

$$|\Psi\rangle = C|\Phi_0\rangle + C_{ai}|\Phi_a^i\rangle + C_{ai,jb}|\Phi_{ab}^{ij}\rangle + \cdots$$

$$|\Phi_0\rangle = \sum_i^M \psi_i = \sum_i^M \sum_k^N c_{ik}\,\phi_k$$

**FIGURE 4.9**   A schematic of the configuration interaction wavefunction and basis set expansion. Here the sum is shown to the second order, and therefore corresponds to configuration interaction singles and doubles (CISD).

possible to simulate these spectra using multiplet theory, an *ab initio* approach is preferable. Toward accounting for these correlation effects in core level spectroscopies, a number of post Hartree–Fock *ab initio* approaches based upon configuration-interaction[137,145–150] and coupled cluster methods[151–154] have been proposed.

For configuration interaction, which we focus upon here, the wavefunction is expressed as the linear combination of Slater determinants (or configurations). The *interaction* refers to the mixing of these different electronic configurations as shown schematically in Figure 4.9. The role of the inclusion of multiconfigurations is to increase the flexibility of the wavefunction. Full configuration interaction, where the excitations are expanded up to all orders is not realistic as the number of determinants grows factorially with the number of electrons in the system, and therefore a truncation of the sum is usually performed. The simplest truncation is configuration interaction singles (CIS), where only single excitations are included. This can be thought of as the wavefunction equivalent to TDDFT within the Tamm–Dancoff approximation, however, due to the success of TDDFT it is now rarely used.

Alternatively, Besley et al. have proposed a method for core excitations based upon CIS(D).[145] This is a second-order perturbative correction to single excitation configuration interaction (CIS). In essence, the double excitations of only CIS active configurations are incorporated[155] and the additional configurations are able to capture many of the modifications of the electronic distribution caused by correlation effects. For valence excited states, CIS(D) often provides a similar accuracy to TDDFT, however, because it includes exact HF exchange, one can expect improvements in excitation energies or states for which charge transfer is important. In a similar vein, Neese and coworkers have recently proposed a combined DFT and restricted open-shell configuration interaction method (DFT/ROCIS).[137,146,147] Their approach expands the reference wavefunction within five excitation classes and includes spin flip excitations through quasi-degenerate perturbation theory using the one electron SOC operator.[144] Importantly, for this hybrid DFT/wavefunction

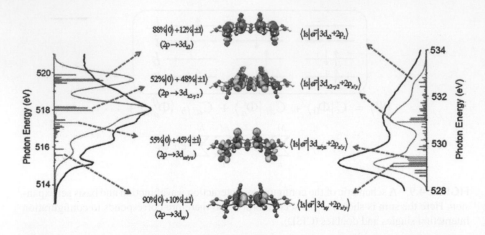

**FIGURE 4.10** (See Color Insert.) Calculated B3LYP/ROCIS V L-edge (red) and O K-edge (green) spectra of the $V_{10}O_{31}H_{12}$ cluster model. The dominant states are assigned in terms of magnetic sublevels, as well as one-electron excitations contributions. The local penta-coordinated environment of vanadium centers is illustrated in the corresponding MO pictures. (Reproduced from Maganas, D. et al., 2013, *Phys. Chem. Chem. Phys.* 15, 7260–7276.[146] With permission from the PCCP Owner Societies.)

approach, a Hartree–Fock wavefunction is a poor choice of reference wavefunction for transition metals as it lacks correlation making it inadequate especially for *d*- and *f*-electrons,[156] which contain a large amount of dynamic correlation. To overcome this, the DFT/ROCIS approach incorporates restricted open-shell Kohn–Sham orbitals into the configuration interaction matrix, following a similar method proposed by Grimme.[157] This introduces a degree of semi-empiricism into the method, however, these optimized parameters are obtained from a fit to a test set of molecules.

The effectiveness of this approach, which can be applied to systems containing more than 100 atoms, is demonstrated in Figure 4.10 showing the V L-edge (red) and O K-edge (green) spectra of $V_2O_5$, which is a potential catalyst involved in the reduction of $NO_x$ by $NH_3$ or in the selective oxidation of hydrocarbons. The DFT/ROCIS approach yields good agreement with both experimental spectra.[146]

Double excitations are usually expected to incorporate ≈95% of the electron correlation. However, at distorted nuclear geometries or when near-degeneracies and multiplets are important, the quality of the wavefunction which incorporates low-order excitations is often not sufficient and this is associated with the inability of CISD to capture the effects of static correlation.[158] In such cases, further flexibility must be incorporated into the wavefunction and this is often achieved by using a multideterminental method, which uses multiple reference wavefunctions with different electronic configurations. An example of such a methodology is the Multi Configuration Self Consistent Field (MCSCF) approach.[159] One important class of MCSCF approach is the Complete Active Space SCF (CASSCF) method.[160] In this scheme the spin orbitals are split into three classes:

- Inactive orbitals: The lower-energy spin orbitals, which remain doubly occupied in all determinants.
- Virtual orbitals: The higher-energy spin orbitals, which remain unoccupied in all determinants.
- Active orbitals: The intermediate energy spin orbitals, which are involved in the configure state functions, which are most important in obtaining an accurate wavefunction. The configuration of the electrons within these active orbitals is calculated from the determinants possible within the active space.

By choosing the most important orbitals, this method allows a complete set of the important determinants to be described accurately, while the reduced configuration space limits the computational expense. The selection of the active space is critical to the accuracy of the calculations and must contain sufficient spin orbitals to describe the appropriate determinants, but keep as small as possible as the number of configure state functions rises rapidly with the number of active orbitals, making calculations harder to execute. It is often said that this approach incorporates the effects of static correlation, while the remaining dynamic correlation must be included using either the MCSCF wavefunction as a reference wavefunction for a perturbative correction (CASPT2)[161,162] or using a configuration interaction (multireference CI), which with a truncation at second-order yields MR-CISD.[150]

Recently, Odelius and coworkers[163,164] have implemented this methodology to calculate core excitations, and by virtue of its high-level treatment of electron correlation effects, they are able to accurately calculate $L_{2/3}$-edge and RIXS spectra. This approach was initially applied to study the $Ni^{2+}(H_2O)_6$ complex as shown in Figure 4.11. This approach is completely *ab initio*, and treats electronically excited states in the same manner as the electronic ground state, making it possible to simulate time-resolved signals used to study changes of the local electronic structure. However, caution must be exercised when using these approaches that the correct configurations are incorporated into the active space. In addition, the computational scaling of this approach remains challenging.

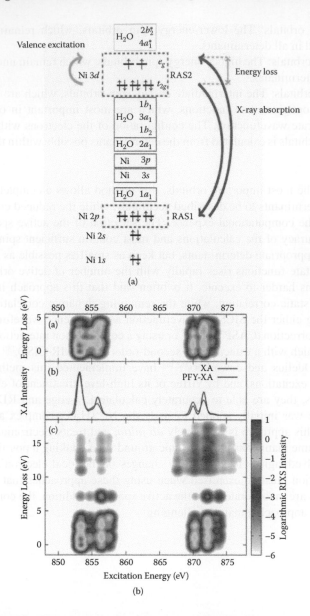

**FIGURE 4.11**   **(See Color Insert.)** Top: Molecular orbital diagram of $Ni^{2+}(H_2O)_6$. For core excitations the active space involves Ni $2p$ and Ni $3d$, partitioned into two subspaces. RAS1 contains the Ni $2p$-orbitals with at most one hole, whereas RAS2 allows all possible electron permutations in the Ni $3d$-orbitals. Ligand-to-metal charge-transfer (LMCT) excitations can be accounted for by introducing ligand orbitals in RAS1 or RAS2. Bottom: (a) The RIXS spectrum corresponding to the RASPT2 X-ray absorption spectrum. (b) The PFY-XA spectrum derived by integrating the RIXS spectrum over all calculated fluorescence decay channels for each excitation energy. (c) RIXS spectrum including charge-transfer (CT) excitations by extending the active space to ligand orbitals. (Figures reproduced from Josefsson, I. et al., 2012, *J. Phys. Chem. Lett.* 3, 3565–3570.[163] With permission.)

# REFERENCES

1. Rehr, J.J. and Albers R.C. Theoretical approaches to X-ray absorption fine structure. *Rev. Mod. Phys.* 2000, 72, 621–654.
2. Natoli, C.R., Misemer, D.K., Doniach, S., and Kutzler, F.W. First-principles calculation of X-ray absorption-edge structure in molecular clusters. *Phys. Rev. A* 1980, 22, 1104–1108.
3. Natoli, C.R., Benfatto, M., and Doniach, S. Use of general potentials in multiple-scattering theory. *Phys. Rev. A* 1986, 34, 4682–4694.
4. de Groot, F.M.F. High-resolution X-ray emission and X-ray absorption spectroscopy. *Chem. Rev.* 2001, 101, 1779–1808.
5. de Groot, F.M.F. Multiplet effects in X-ray spectroscopy. *Coordin. Chem. Rev.* 2005, 249, 31–63.
6. Stavitski, E. and de Groot, F.M.F. The CTM4XAS program for EELS and XAS spectral shape analysis of transition metal L edges. *Micron.* 2010, 41, 687–694.
7. de Groot, F.M.F. and Kotani, A. *Core Level Spectroscopy of Solides*, Boca Raton, FL: CRC Press, 2008.
8. de Groot, F.M.F., Fuggle, J.C., Thole, B.T., and Sawatzky, G.A. A. 2p X-ray absorption of 3d transition-metal compounds: An atomic multiplet description including the crystal field. *Phys. Rev. B* 1990, 42, 5459–5468.
9. Glatzel, P. and Bergmann, U. High resolution 1s core hole X-ray spectroscopy in 3d transition metal complexes—Electronic and structural information. *Coordin. Chem. Rev.* 2005, 249, 65–95.
10. Lima, F.A., Milne, C.J., Amarasinghe, D.C.V., Rittmann-Frank, M.H., Van der Veen, R.M., Reinhard, M., Pham, V.T., Karlsson, S., Johnson, S.L., Grolimund, D., Borca, C., Huthwelker, T., Janousch, M., Van Mourik, F., Abela, R., and Chergui, M. A high-repetition rate scheme for synchrotron-based picosecond laser pump/X-ray probe experiments on chemical and biological systems in solution. *Rev. Sci. Instrum.* 2011, 82, 063111.
11. Pham, V.T., Penfold, T.J., Van der Veen, R.M., Lima, F.A., El Nahhas, A., Johnson, S.L., Beaud, P., Abela, R., Bressler, C., Tavernelli, I., Milne, C.J., and Chergui, M. Probing the transition from hydrophilic to hydrophobic solvation with atomic scale resolution. *J. Am. Chem. Soc.* 2011, 133, 12740–12748.
12. Bressler, C., Milne, C.J., Pham, V.T., El Nahhas, A., Van der Veen, R.M., Gawelda, W., Johnson, S.L., Beaud, P., Grolimund, D., Kaiser, M., Borca, C., Ingold, G., Abela, R., and Chergui, M. Femtosecond XANES study of the light-induced spin crossover dynamics in an iron(II) complex. *Science* 2009, 323, 489–492.
13. Huse, N., Cho, H., Hong, K., Jamula, L., de Groot, F.M.F., Kim, T.K., McCusker, J.K., and Schoenlein, R.W. Femtosecond soft X-ray spectroscopy of solvated transition-metal complexes: Deciphering the interplay of electronic and structural dynamics. *J. Phys. Chem. Lett.* 2011, 2, 880–884.
14. Haldrup, K., Vankó, G., Gawelda, W., Galler, A., Doumy, G., March, A.M., Kanter, E.P., Bordage, A., Dohn, A., Van Driel, T.B., Kjaer, K.S., Lemke, H.T., Canton, S.E., Uhlig, J., Sundstrom, V., Young, L., Southworth, S.H., Nielsen, M.M., and Bressler, C. Guest-host interactions investigated by time-resolved X-ray spectroscopies and scattering at MHz rates: Solvation dynamics and photo-induced spin transition in aqueous [Fe(bipy)₃]²⁺. *J. Phys. Chem. A* 2012, 116, 9878–9887.
15. Vanko, G., Bordage, A., Glatzel, P., Gallo, E., Rovezzi, M., Gawelda, W., Galler, A., Bressler, C., Doumy, G., March, A.M., Kanter, E.P., Young, L., Southworth, S.H., Canton, S.E., Uhlig, J., Smolentsev, G., Sundström, V., Haldrup, K., Van Driel, T.B., Nielsen, M.M., Kjaer, K.S., and Lemke, H.T. Spin-state studies with XES and RIXS: From static to ultrafast. *Journal of Electron Spectroscopy and Related Phenomena* 2013, 188, 166–171, http://dx.doi.org/10.1016/j.elspec.2012.09.012.

16. Khan, S. Free-electron lasers. *J. Mod. Optics* 2008, 55, 3469–3512.
17. Mukamel, S., Healion, D., Zhang, Y., and Biggs, J.D. Multidimensional attosecond resonant X-ray spectroscopy of molecules: Lessons from the optical regime. *Annu. Rev. Phys. Chem.* 2013, 64, 101–127.
18. Bunker, G. *Introduction to XAFS: A Practical Guide to X-ray Absorption Fine Structure Spectroscopy*, 1st ed. Cambridge: Cambridge University Press, 2010.
19. Stöhr, J. *NEXAFS Spectroscopy*, 1st ed. Heidelberg: Springer, 1992.
20. Ament, L.J.P., Van Veenendaal, M., Devereaux, T.P., Hill, J.P., and Van den Brink, Resonant inelastic X-ray scattering studies of elementary excitations. *J. Rev. Mod. Phys.* 2011, 83, 705–767.
21. Rehr, J.J., Kas, J.J., Prange, M.P., Sorini, A.P., Takimoto, Y., and Vila, F. *Ab initio* theory and calculations of X-ray spectra. *Comptes Rendus Physique* 2009, 10, 548–559.
22. Rehr, J.J. and Ankudinov, A.L. A. Progress in the theory and interpretation of XANES. *Coordin. Chem. Rev.* 2005, 249, 131–140.
23. Bressler, C., Abela, R., and Chergui, M. Exploiting EXAFS and XANES for time-resolved molecular structures in liquids. *Z. Kristallogr.* 2008, 223, 307–321.
24. Penfold, T.J., Milne, C.J., and Chergui, M. Recent advances in ultrafast X-ray absorption spectroscopy of solutions. *Adv. Chem. Phys.* 2013, 153, 1–44.
25. Chen, L.X. Probing transient molecular structures in photochemical processes using laser-initiated time-resolved X-ray absorption spectroscopy. *Annu. Rev. Phys. Chem.* 2005, 56, 221–254.
26. Hämäläinen, K. and Manninen, S. Resonant and non-resonant inelastic x-ray scattering. *J. Phys-Condens. Mat.* 2001, 13, 7539–7556.
27. Debeer-George, S., Petrenko, T., and Neese, F. Time-dependent density functional calculations of ligand K-edge X-ray absorption spectra. *Inorg. Chim. Acta* 2008, 361, 965–972.
28. Bernadotte, S., Atkins, A.J., and Jacob, C.R. Origin-independent calculation of quadrupole intensities in X-ray spectroscopy. *J. Chem. Phys.* 2012, 137, 204106.
29. Westre, T.E., Kennepohl, P., DeWitt, J.G., Hedman, B., Hodgson, K.O., and Solomon, E.I. A multiplet analysis of Fe K-edge 1s → 3d pre-edge features of iron complexes. *J. Am. Chem. Soc.* 1997, 119, 6297–6314.
30. Hämäläinen, K., Siddons, D.P., Hastings, J.B., and Berman, L.E Elimination of the inner-shell lifetime broadening in X-ray-absorption spectroscopy. *Phys. Rev. Lett.* 1991, 67, 2850–2853.
31. Hämäläinen, K., Kao, C.C., Hastings, J.B., Siddons, D.P., Berman, L.E., Stojanoff, V., and Cramer, S.P. Spin-dependent X-ray absorption of MnO and MnF$_2$. *Phys. Rev. B* 1992, 46, 14274–14277.
32. Kotani, A. and Shin, S. Resonant inelastic X-ray scattering spectra for electrons in solids. *Rev. Mod. Phys.* 2001, 73, 203–246.
33. Gel'mukhanov, F. and Agren, H. Resonant X-ray Raman scattering. *Physics Reports* 1999, 312, 87–330.
34. Kotani, A., Kvashnina, K.O., Butorin, S.M., and Glatzel, P. Spectator and participator processes in the resonant photon-in and photon-out spectra at the Ce L$_3$ edge of CeO$_2$. *Eur. Phys. J. B* 2012, 85, 257.
35. Glatzel, P., Sikora, M., and Fernández-García, M. Resonant X-ray spectroscopy to study K absorption pre-edges in 3d transition metal compounds. *Eur. Phys. J. Spec. Top.* 2009, 169, 207–214.
36. Lee, P., Citrin, P., Eisenberger, P., and Kincaid, B. Extended X-ray absorption fine structure—Its strengths and limitations as a structural tool. *Rev. Mod. Phys.* 1981, 53, 769–806.
37. Köningsberger, D., Mojet, B., Van Dorssen, G., and Ramaker, D. XAFS spectroscopy: Fundamental principles and data analysis. *Top Catalysis* 2000, 10, 143–155.

38. Weng, T.C., Waldo, G.S., and Penner-Hahn, J.E. A method for normalization of X-ray absorption spectra. *J. Synchrotron Radiat.* 2005, 12, 506–510.
39. Ravel, B. and Newville, M. ATHENA, ARTEMIS, HEPHAESTUS: Data analysis for X-ray absorption spectroscopy using IFEFFIT. *J. Synchrotron Radiat.* 2005, 12, 537–541.
40. Lima, F.A., Penfold, T.J., Van der Veen, R.M., Reinhard, M., Abela, R., Tavernelli, I., Rothlisberger, U., Benfatto, M., Milne, C.J., and Chergui, M. Probing the electronic and geometric structure of ferric and ferrous myoglobins in physiological solutions by Fe K-edge absorption spectroscopy. *Phys. Chem. Chem. Phys.* 2014, 16, 1617–1631.
41. Rehr, J., de Leon, J.M., Zabinsky, S.J., and Albers, R.C. Theoretical X-ray absorption fine structure standards. *J. Am. Chem. Soc.* 1991, 113, 5135–5140.
42. Newville, M., Livins, P., Yacoby, Y., Rehr, J.J., and Stern, E. Near-edge X-ray-absorption fine-structure of Pb—A comparison of theory and experiment. *Phys. Rev. B* 1993, 47, 14126–14131.
43. Als-Nielsen, J. and McMorrow, D. *Elements of Modern X-ray Physics.* New York: Wiley, 2011.
44. Jaklevic, J., Kirby, J.A., Klein, M.P., Robertson, A.S., Brown, G.S., and Eisenberger, P. Fluorescence detection of EXAFS: Sensitivity enhancement for dilute species and thin films. *Solid State Communications* 1977, 23, 679–682.
45. de Groot, F.M.F., Arrio, M.A., Sainctavit, P., Cartier, C., and Chen, C.T. Fluorescence yield detection: Why it does not measure the X-ray absorption cross section. *Solid State Communications* 1994, 92, 991–995.
46. Booth, C.H. and Bridges, F. Improved self-absorption correction for fluorescence measurements of extended X-ray absorption fine structure. *Phys. Scripta* 2005, 202.
47. Dallera, C., Braicovich, L., Ghiringhelli, G., Van Veenendaal, M., Goedkoop, J., and Brookes, N. Resonant soft-X-ray inelastic scattering from Gd in the $Gd_3Ga_5O_{12}$ garnet with excitation across the $M_5$ edge. *Phys. Rev. B* 1997, 56, 1279–1283.
48. Chabot-Couture, G., Hancock, J.N., Mang, P.K., Casa, D.M., Gog, T., and Greven, M. Polarization dependence and symmetry analysis in indirect K-edge RIXS. *Phys. Rev. B* 2010, 82, 035113.
49. Ankudinov, A. Rehr, J. Sum rules for polarization-dependent X-ray absorption. *J. Phys. Rev. B* 1995, 51, 1282–1285.
50. Hocking, R.K., Wasinger, E.C., de Groot, F.M.F., Hodgson, K.O., Hedman, B., and Solomon, E.I. Fe L-edge XAS studies of $K_4[Fe(CN)_6]$ and $K_3[Fe(CN)_6]$: A direct probe of back-bonding. *J. Am. Chem. Soc.* 2006, 128, 10442–10451.
51. Carra, P., König, H., Thole, B.T., and Altarelli, M. Magnetic X-ray dichroism: General features of dipolar and quadrupolar spectra. *Physica B: Physics of Condensed Matter* 1993, 192, 182–190.
52. Thole, B.T., Carra, P., Sette, F., and Van der Laan, G. X-ray circular dichroism as a probe of orbital magnetization. *Phys. Rev. Lett.* 1992, 68, 1943–1946.
53. Altarelli, M. Orbital-magnetization sum rule for X-ray circular dichroism: A simple proof. *Phys. Rev. B* 1993, 47, 597–598.
54. Carra, P., Fabrizio, M., and Thole, B.T. High resolution X-ray resonant Raman scattering. *Phys. Rev. Lett.* 1995, 74, 3700–3703.
55. Van der Veen, R.M., Cannizzo, A., Van Mourik, F., Vlček, A., and Chergui, M. Vibrational relaxation and intersystem crossing of binuclear metal complexes in solution. *J. Am. Chem. Soc.* 2011, 133, 305–315.
56. Penfold, T., Karlsson, S., Capano, G., Lima, F.A., Rittmann, J., Reinhard, M., Rittmann-Frank, M.H., Braem, O., Baranoff, E., Abela, R., Tavernelli, I., Rothlisberger, U., Milne, C.J., and Chergui, M. Solvent-induced luminescence quenching: Static and time-resolved X-ray absorption spectroscopy of a copper (I) phenanthroline complex. *J. Phys. Chem. A* 2013, 117, 4591–4601.

57. Bunker, G.B. A X-Ray Absorption Study of Transition Metal Oxides, Ph.D. Thesis, Washington: Univ. of Washington, 1984.

58. Webb, S.M. SIXpack: A graphical user interface for XAS analysis using IFEFFIT. *Phys. Scripta* 2005, T115, 1011–1014.

59. Smolentsev, G. and Soldatov, A. Quantitative local structure refinement from XANES: Multi-dimensional interpolation approach. *J. Synchrotron Radiat.* 2006, 13, 19–29.

60. Benfatto, M., Congiu-Castellano, A., Daniele, A., and Della-Longa, S. MXAN: A new software procedure to perform geometrical fitting of experimental XANES spectra. *J. Synchrotron Radiat.* 2001, 8, 267–269.

61. Joly, Y. X-ray absorption near-edge structure calculations beyond the muffin-tin approximation. *Phys. Rev. B* 2001, 63, 1–10.

62. Hermann, K., Pettersson, L.G.M., Casida, M.E., Daul, C., Goursot, A., Koester, A., Proynov, E., St-Amant, A., Salahub, D.R., StoBe-deMon version 3.2 (2013), http://www.fhi-berlin.mpg.de/KHsoftware/StoBe/index.html.

63. Neese, F. The ORCA program system. *Wiley Interdisciplinary Reviews: Computational Molecular Science* 2011, 2, 73–78.

64. te Velde, G., Bickelhaupt, F.M., Baerends, E.J., Fonseca Guerra, C., Van Gisbergen, S.J., Snijders, J.G., and Ziegler, T. Chemistry with ADF. *J. Comput. Chem.* 2001, 22, 931–967.

65. Shao, Y., Molnar, L.F., Jung, Y., Kussmann, J.R., Ochsenfeld, C., Brown, S.T., Gilbert, A.T.B., Slipchenko, L.V., Levchenko, S.V., O Neill, D.P., DiStasio, R.A., Jr., Lochan, R.C., Wang, T., Beran, G.J.O., Besley, N.A., Herbert, J.M., Yeh Lin, C., Van Voorhis, T., Hung Chien, S., Sodt, A., Steele, R.P., Rassolov, V.A., Maslen, P.E., Korambath, P.P., Adamson, R.D., Austin, B., Baker, J., Byrd, E.F.C., Dachsel, H., Doerksen, R.J., Dreuw, A., Dunietz, B.D., Dutoi, A.D., Furlani, T.R., Gwaltney, S.R., Heyden, A., Hirata, S., Hsu, C.-P., Kedziora, G., Khalliulin, R.Z., Klunzinger, P., Lee, A.M., Lee, M.S., Liang, W., Lotan, I., Nair, N., Peters, B., Proynov, E.I., Pieniazek, P.A., Min Rhee, Y., Ritchie, J., Rosta, E., David Sherrill, C., Simmonett, A.C., Subotnik, J.E., Lee Woodcock, H., III., Zhang, W., Bell, A.T., Chakraborty, A.K., Chipman, D.M., Keil, F.J., Warshel, A., Hehre, W.J., Schaefer, H.F., III., Kong, J., Krylov, A.I., Gill, P.M.W., and Head-Gordon, M. Advances in methods and algorithms in a modern quantum chemistry program package. *Phys. Chem. Chem. Phys* 2006, 8, 3172.

66. Valiev, M., Bylaska, E.J., Govind, N., Kowalski, K., Straatsma, T.P., Van Dam, H.J., Wang, D., Nieplocha, J., Apra, E., and Windus, T.L. NWChem: A comprehensive and scalable open-source solution for large-scale molecular simulations. *Computer Physics Communications* 2010, 181, 1477–1489.

67. Schwarz, K. and Blaha, P. Electronic structure of solids with WIEN2k. *Mol Phys* 2010.

68. Gonze, X., Amadon, B., Anglade, P.M., Beuken, J.M., Bottin, F., Boulanger, P., Bruneval, F., Caliste, D., Caracas, R., Côté, M., Deutsch, T., Genovese, L., Ghosez, P., Giantomassi, M., Goedecker, S., Hamann, D.R., Hermet, P., Jollet, F., Jomard, G., Leroux, S., Mancini, M., Mazevet, S., Oliveira, M.J.T., Onida, G., Pouillon, Y., Rangel, T., Rignanese, G.M., Sangalli, D., Shaltaf, R., Torrent, M., Verstraete, M.J., Zerah, G., and Zwanziger, J.W. ABINIT: First-principles approach to material and nanosystem properties. *Computer Physics Communications* 2009, 180, 2582–2615.

69. Lippert, G., Hutter, J., and Parrinello, M. The Gaussian and augmented-plane-wave density functional method for *ab initio* molecular dynamics simulations. *Theor. Chem. Acc* 1999, 103, 124–140.

70. Campbell, L., Hedin, L., Rehr, J.J., and Bardyszewski, W. Interference between extrinsic and intrinsic losses in X-ray absorption fine structure. *Phys. Rev. B* 2002, 65, 064107.

71. Papanikolaou, N., Zeller, R., and Dederichs, P.H. Conceptual improvements of the KKR method. *J. Phys-Condens Mat.* 2002, 14, 2799–2823.

72. Hatada, K., Hayakawa, K., Benfatto, M., and Natoli, C.R. Full-potential multiple scattering theory with space-filling cells for bound and continuum states. *J. Phys-Condens Mat.* 2010, 22, 185501.

73. Gonis, A. and Butler, W.H. *Multiple Scattering in Solids.* New York: Springer, 2000.

74. Williams, A.R. and Morgan, J.V.W. Multiple scattering by non-muffin-tin potentials: General formulation. *Journal of Physics C: Solid State Physics* 1974, 7, 37.

75. Natoli, C.R., Krüger, P., Hatada, K., Hayakawa, K., Sébilleau, D., and Šipr, O., Multiple scattering theory for non-local and multichannel potentials. *J. Phys-Condens. Mat.* 2012, 24, 365501.

76. Wende, H., Srivastava, P., Chauvistré, R., May, F., Baberschke, K., Arvanitis, D., and Rehr, J.J. Evidence for photoelectron backscattering by interstitial charge densities. *J. Phys-Condens. Mat.* 1999, 9, L427–L433.

77. Hedin, L. and Lundqvist, B.I. Explicit local exchange-correlation potentials. *Journal of Physics C: Solid State Physics* 1971, 4, 2064–2083.

78. Hedin, L., Lundqvist, B.I., and Lundqvist, S. Local exchange-correlation potentials. *Solid State Communications* 1971, 9, 537–541.

79. Natoli, C.R., Benfatto, M., Brouder, C., Ruiz Lopez, M., and Foulis, D. Multichannel multiple-scattering theory with general potentials. *Phys. Rev. B* 1990, 42, 1944–1968.

80. Kas, J.J., Sorini, A., Prange, M., Campbell, L., Soininen, J., and Rehr, J.J. Many-pole model of inelastic losses in X-ray absorption spectra. *Phys. Rev. B* 2007, 76, 195116.

81. Rehr, J.J., Soininen, J.A., and Shirley, E.L. Final-state rule vs. the Bethe-Salpeter equation for deep-core X-ray absorption spectra. *Physica Scripta.* 2005, T115, 207–211.

82. Briois, V., Sainctavit, P., Long, G.J., and Grandjean, F. Importance of photoelectron multiple scattering in the iron K-edge X-ray absorption spectra of spin-crossover complexes: Full multiple scattering calculations for several iron(II) tyrispyrazolylborate and trispyrazolylmethane complexes. *Inorg. Chem.* 2001, 40, 912–918.

83. El Nahhas, A., Van der Veen, R.M., Penfold, T.J., Pham, V.T., Lima, F.A., Abela, R., Blanco-Rodriguez, A.M., Zális, S., Vlček, A., Tavernelli, I., Rothlisberger, U., Milne, C.J., and Chergui, M. X-ray absorption spectroscopy of ground and excited rhenium–carbonyl–diimine complexes: Evidence for a two-center electron transfer. *J. Phys. Chem. A* 2013, 117, 361–369.

84. Van der Veen, R.M., Kas, J.J., Milne, C.J., Pham, V.T., El Nahhas, A., Lima, F.A., Vithanage, D.A., Rehr, J.J., Abela, R., and Chergui, M. L-edge XANES analysis of photo-excited metal complexes in solution. *Phys. Chem. Chem. Phys.* 2010, 12, 5551–5561.

85. Danese, J.B. Calculation of the total energy in the multiple scattering-Xα method. II. *J. Chem. Phys.* 1974, 61, 3071.

86. Danese, J.B. and Connolly, J. Calculation of the total energy in the multiple scattering-Xα method. I. *J. Chem. Phys.* 1974, 61, 3063.

87. Joly, Y. Finite-difference method for the calculation of low-energy positron diffraction. *Phys. Rev. B* 1996, 53, 13029.

88. Gawelda, W., Pham, V.T., Benfatto, M., Zaushitsyn, Y., Kaiser, M., Grolimund, D., Johnson, S.L., Abela, R., Hauser, A., Bressler, C., and Chergui, M. Structural determination of a short-lived excited iron(II) complex by picosecond X-ray absorption spectroscopy. *Phys. Rev. Lett.* 2007, 98, 057401.

89. Gawelda, W., Pham, V.T., Van der Veen, R.M., Grolimund, D., Abela, R., Chergui, M., and Bressler, C. Structural analysis of ultrafast extended X-ray absorption fine structure with subpicometer spatial resolution: Application to spin crossover complexes. *J. Chem. Phys.* 2009, 130, 124520.

90. Chen, L.X., Shaw, G., Novozhilova, I., Liu, T., Jennings, G., Attenkofer, K., Meyer, G., and Coppens, P. MLCT state structure and dynamics of a copper(I) diimine complex characterized by pump-probe X-ray and laser spectroscopies and DFT calculations. *J. Am. Chem. Soc.* 2003, 125, 7022–7034.

91. Smolentsev, G., Soldatov, A., and Chen, L.X. Three-dimensional local structure of photoexcited Cu diimine complex refined by quantitative XANES analysis. *J. Phys. Chem. A* 2008, 112, 5363–5367.

92. Witte, C., Chantler, C., Cosgriff, E. and Tran, C. Atomic cluster calculation of the X-ray near-edge absorption of copper. *Radiation Physics and Chemistry* 2006, 75, 1582–1585.

93. Blöchl, P.E. Projector augmented-wave method. *Phys. Rev. B* 1994, 50, 17953.

94. Taillefumier, M., Cabaret, D., Flank, A.M. and Mauri, F. X-ray absorption near-edge structure calculations with the pseudopotentials: Application to the K edge in diamond and α-quartz. *Phys. Rev. B* 2002, 66, 195107.

95. Bunau, O. and Calandra, M. Projector augmented wave calculation of X-ray absorption spectra at the $L_{2,3}$ edges. *Phys. Rev. B* 2013, 87, 205015.

96. Van de Vondele, J., Iannuzzi, M., and Hutter, J. Large-scale condensed matter calculations using the Gaussian and augmented plane waves method. *Lecture Notes in Physics* 2006, 703, 297–314.

97. Iannuzzi, M. and Hutter, J. Inner-shell spectroscopy by the Gaussian and augmented plane wave method. *PCCP* 2007, 9, 1599–1610.

98. Iannuzzi, M. X-ray absorption spectra of hexagonal ice and liquid water by all-electron Gaussian and augmented plane wave calculations. *J. Chem. Phys.* 2008, 128, 204506.

99. Goulon, J., Goulon-Ginet, C., Cortes, R., and Dubois, J.M. On experimental attenuation factors of the amplitude of the EXAFS oscillations in absorption, reflectivity and luminescence measurements. *J. Phys. France* 1982, 43, 539–548.

100. Glatzel, P., Jacquamet, L., Bergmann, U., de Groot, F.M.F., and Cramer, S.P. Site-selective EXAFS in mixed-valence compounds using high-resolution fluorescence detection: A study of iron in Prussian blue. *Inorg. Chem.* 2002, 41, 3121–3127.

101. Yano, J., Pushkar, Y., Glatzel, P., Lewis, A., Sauer, K., Messinger, J., Bergmann, U., and Yachandra, V. High-resolution Mn EXAFS of the oxygen-evolving complex in photosystem II: Structural implications for the $Mn_4Ca$ cluster. *J. Am. Chem. Soc.* 2005, 127, 14974–14975.

102. Wang, X., Randall, C.R., Peng, G., and Cramer, S.P. Spin-polarized and site-selective X-ray absorption. Demonstration with Fe porphyrins and Kβ detection. *Chem. Phys. Lett.* 1995, 243, 469–473.

103. Zabinsky, S., Rehr, J.J., Ankudinov, A., and Albers, R.C. Multiple-scattering calculations of X-ray-absorption spectra. *Phys. Rev. B* 1995, 52, 2995–3009.

104. Sevillano, E., Meuth, H., and Rehr, J. Extended X-ray absorption fine structure Debye–Waller factors. I. Monatomic crystals. *J. Phys. Rev. B* 1979, 20, 4908–4911.

105. Poiarkova, A.V. and Rehr, J.J. Multiple-scattering X-ray-absorption fine-structure Debye–Waller factor calculations. *Phys. Rev. B* 1999, 59, 948.

106. Provost, K., Beret, E.C., Muller, D.B., Michalowicz, A., and Marcos, E.S. EXAFS Debye–Waller factors issued from Car–Parrinello molecular dynamics: Application to the fit of oxaliplatin and derivatives. *J. Chem. Phys.* 2013, 138, 084303.

107. Sayers, D.E., Stern, E.A., and Lytle, F.W. New technique for investigating noncrystalline structures: Fourier analysis of the extended X-ray absorption fine structure. *Phys. Rev. Lett.* 1971, 27, 1204–1207.

108. Frenkel, A.I. Applications of extended X-ray absorption fine-structure spectroscopy to studies of bimetallic nanoparticle catalysts. *Chem. Soc. Rev.* 2012, 41, 8163.

109. Alayoglu, S., Zavalij, P., Eichhorn, B., Wang, Q., Frenkel, A.I., and Chupas, P. Structural and architectural evaluation of bimetallic nanoparticles: A case study of Pt–Ru core–shell and alloy nanoparticles. *ACS Nano* 2009, 3, 3127–3137.

110. Riggs-Gelasco, P.J., Stemmler, T.L., and Penner-Hahn, J.E. XAFS of dinuclear metal sites in proteins and model compounds. *Coordin. Chem. Rev.* 1995, 144, 245–286.

111. Funke, H., Chukalina, M., and Scheinost, A.C. A new FEFF-based wavelet for EXAFS data analysis. *J. Synchrotron Radiat.* 2007, 14, 426–432.

112. Funke, H., Scheinost, A., and Chukalina, M. Wavelet analysis of extended X-ray absorption fine structure data. *Phys. Rev. B* 2005, 71.

113. Penfold, T.J., Tavernelli, I., Milne, C.J., Reinhard, M., El Nahhas, A., Abela, R., Rothlisberger, U., and Chergui, M. A wavelet analysis for the X-ray absorption spectra of molecules. *J. Chem. Phys.* 2013, 138, 014104.

114. Stern, E.A. Number of relevant independent points in X-ray-absorption fine-structure spectra. *Phys. Rev. B* 1993, 48, 9825–9827.

115. DeBeer-George, S., Petrenko, T., and Neese, F. Prediction of iron K-edge absorption spectrausing time-dependent density functional theory. *J. Phys. Chem. A* 2008, 112, 12936–12943.

116. Roemelt, M., Beckwith, M.A., Duboc, C., Collomb, M.N., Neese, F., and DeBeer, S. Manganese K-edge X-ray absorption spectroscopy as a probe of the metal–Ligand interactions in coordination compounds. *Inorg. Chem.* 2012, 51, 680–687.

117. Atkins, A.J., Bauer, M., and Jacob, C.R. The chemical sensitivity of X-ray spectroscopy: High energy resolution XANES versus X-ray emission spectroscopy of substituted ferrocenes. *Phys. Chem. Chem. Phys.* 2013,15, 8095–8105.

118. Besley, N.A. and Asmuruf, F.A. Time-dependent density functional theory calculations of the spectroscopy of core electrons. *Phys. Chem. Chem. Phys.* 2010, 12, 12024.

119. Stener, M. Time-dependent density functional theory of core electrons excitations. *Chem. Phys. Lett.* 2003, 373, 115–123.

120. Ray, K., Debeer G.S., Solomon, E.I., Wieghardt, K., Neese, F. Description of the ground-state covalencies of the bis(dithiolato) transition-metal complexes from X-ray absorption spectroscopy and time-dependent density-functional calculations. *Chem. Eur. J.* 2007, 13, 2783–2797.

121. Marques, M.A.L. and Gross, E.K.U. Time-dependent density functional theory. *Annu Rev. Phys. Chem.* 2004, 55, 427–455.

122. Ullrich, C.A. *Time-Dependent Density-Functional Theory: Concepts and Applications.* New York: Oxford University Press, 2011.

123. Lopata, K., Van Kuiken, B.E., Khalil, M., and Govind, N. Time-dependent density functional theory studies of core-level near-edge X-ray absorption. *J. Chem. Theory Comput.* 2012, 8, 3284–3292.

124. Casida, M.E. In *Recent Advances in Density Functional Methods, Vol. 1,* Ed. Chong, D.P. Singapore: World Scientific, 1995.

125. Casida, M. and Huix-Rotllant, M. Progress in time-dependent density-functional theory. *Annu. Rev. Phys. Chem.* 2012, 63, 287–323.

126. Hirata, S. and Head-Gordon, M. Time-dependent density functional theory within the Tamm–Dancoff approximation. *Chem. Phys. Lett.* 1999, 314, 291–299.

127. Asmuruf, F.A. and Besley, N.A. Time-dependent density functional theory study of the near-edge X-ray absorption fine structure of benzene in gas phase and on metal surfaces. *J. Chem. Phys.* 2008, 129, 064705.

128. Perdew, J. Density-functional approximation for the correlation-energy of the inhomogeneous electron-gas. *Phys. Rev. B* 1986, 33, 8822–8824.

129. Becke, A.D. Density-functional exchange-energy approximation with correct asymptotic behavior. *Phys. Rev. A* 1988, 38, 3098.
130. Perdew, J.P., Burke, K., and Ernzerhof, M. Generalized gradient approximation made simple. *Phys. Rev. Lett.* 1996, 77, 3865–3868.
131. Stephens, P.J., Devlin, F.J., Chabalowski, C.F., and Frisch, M.J. *Ab initio* calculation of vibrational absorption and circular dichroism spectra using density functional force fields. *J. Phys. Chem.* 1994, 98, 11623–11627.
132. Ernzerhof, M. and Scuseria, G.E. Assessment of the Perdew–Burke–Ernzerhof exchange-correlation functional. *J. Chem. Phys.* 1999, 110, 5029–5036.
133. Van Leeuwen, R. and Baerends, E.J. Exchange-correlation potential with correct asymptotic behavior. *Phys. Rev. A* 1994, 49, 2421–2431.
134. Perdew, J.P. Self-interaction correction to density-functional approximations for many-electron systems. *Phys. Rev. B* 1981, 23, 5048–5079.
135. Casida, M., Jamorski, C., Casida, K.C., and Salahub, D.R. Molecular excitation energies to high-lying bound states from time-dependent density-functional response theory: Characterization and correction of the time-dependent local density approximation ionization threshold. *J. Chem. Phys.* 1998, 108, 4439.
136. Debeer-George, S. and Neese, F. Calibration of scalar relativistic density functional theory for the calculation of sulfur K-edge X-ray absorption spectra. *Inorg. Chem.* 2010, 49, 1849–1853.
137. Roemelt, M. and Neese, F. Excited states of large open-shell molecules: An efficient, general, and spin-adapted approach based on a restricted open-shell ground state wave function. *J. Phys. Chem. A* 2013, 117, 3069–3083.
138. Capano, G., Penfold, T.J., Besley, N.A., Milne, C.J., Reinhard, M., Rittmann-Frank, H., Glatzel, P., Abela, R., Rothlisberger, U., Chergui, M., and Tavernelli, I. The role of Hartree–Fock exchange in the simulation of X-ray absorption spectra: A study of photoexcited $[Fe(bpy)_3]^{2+}$.*Chem. Phys. Lett.* 2013, 580, 179–184, http://dx.doi.org/10.1016/j.cplett.2013.06.060.
139. Dreuw, A., Weisman, J.L., and Head-Gordon, M. Long-range charge-transfer excited states in time-dependent density functional theory require non-local exchange. *J. Chem. Phys.* 2003, 119, 2943–2946.
140. de Groot, F.M.F., Hu, Z.W., Lopez, M.F., Kaindl, G., Guillot, F., and Tronc, M. Differences between $L_3$ and $L_2$ X-ray absorption spectra of transition metal compounds. *J. Chem. Phys* 1994, 101, 6570.
141. Cowan, R.D. *The Theory of Atomic Structure and Spectra.* Berkeley: University of California Press, 1981.
142. Gawelda, W., Johnson, M., de Groot, F.M.F., Abela, R., Bressler, C., and Chergui, M. Electronic and molecular structure of photoexcited $[Ru(II)(bpy)_3]^{2+}$ probed by picosecond X-ray absorption spectroscopy. *J. Am. Chem. Soc.* 2006, 128, 5001–5009.
143. Ikeno, H., de Groot, F.M.F., and Tanaka, I. *Ab-initio* CI calculations for 3d transition metal $L_{2,3}$ X-ray absorption spectra of TiCl4 and VOCl3. *J. Phys.: Conf. Ser.* 2009, 190, 012005.
144. Neese, F., Petrenko, T., Ganyushin, D., and Olbrich, G. Advanced aspects of *ab initio* theoretical optical spectroscopy of transition metal complexes: Multiplets, spin-orbit coupling and resonance Raman intensities. *Coordin Chem. Rev* 2007, 251, 288–327.
145. Asmuruf, F.A. and Besley, N.A. Calculation of near-edge X-ray absorption fine structure with the CIS(D) method. *Chem. Phys. Lett.* 2008, 463, 267–271.
146. Maganas, D., Roemelt, M., Hävecker, M., Trunschke, A., Knop-Gericke, A., Schlögl, R., and Neese, F. First principles calculations of the structure and V L-edge X-ray absorption spectra of V2O5 using local pair natural orbital coupled cluster theory and spin-orbit coupled configuration interaction approaches. *Phys. Chem. Chem. Phys.* 2013, 15, 7260–76.

147. Roemelt, M., Maganas, D., DeBeer, S., and Neese, F. A combined DFT and restricted open-shell configuration interaction method including spin-orbit coupling: Application to transition metal L-edge X-ray absorption spectroscopy. *J. Chem. Phys.* 2013, 138, 204101.

148. Ikeno, H., de Groot, F.M.F., Stavitski, E., and Tanaka, I. Multiplet calculations of $L_{2,3}$ X-ray absorption near-edge structures for 3d transition-metal compounds. *J. Phys- Condens. Mat.* 2009, 21, 104208.

149. Ikeno, H., Tanaka, I., Koyama, Y., Mizoguchi, T., and Ogasawara, K. First-principles multielectron calculations of Ni $L_{2,3}$ NEXAFS and ELNES for LiNiO2 and related compounds. *Phys. Rev. B* 2005, 72, 075123.

150. Brabec, J., Bhaskaran-Nair, K., Govind, N., Pittner, J., and Kowalski, K. Application of state-specific multireference coupled cluster methods to core-level excitations. *J. Chem. Phys.* 2012, 137, 171101.

151. Nooijen, M. and Bartlett, R.J. Description of core-excitation spectra by the open-shell electron-attachment equation of motion coupled cluster method. *J. Chem. Phys.* 1995, 102, 6735.

152. Coriani, S., Christiansen, O., Fransson, T., and Norman, P. Coupled-cluster response theory for near-edge X-ray-absorption fine structure of atoms and molecules. *Phys. Rev. A* 2012, 85, 022507.

153. Fransson, T., Coriani, S., Christiansen, O., and Norman, P. A study at the coupled cluster, density functional, and static-exchange levels of theory. *J. Chem. Phys.* 2013, 138, 124311.

154. Besley, N.A. Equation of motion coupled cluster theory calculations of the X-ray emission spectroscopy of water. *Chem. Phys. Lett.* 2012, 542, 42–46.

155. Head-Gordon, M., Rico, R.J., Oumi, M., and Lee, T.J. A doubles correction to electronic excited states from configuration interaction in the space of single substitutions. *Chem. Phys. Lett.* 1994, 219, 21–29.

156. Furche, F. and Perdew, J.P. The performance of semilocal and hybrid density functionals in 3d transition-metal chemistry. *J. Chem. Phys.* 2006, 124, 044103.

157. Grimme, S. and Waletzke, M. A combination of Kohn–Sham density functional theory and multi-reference configuration interaction methods. *J. Chem. Phys.* 1999, 111, 5645–5655.

158. Tew, D.P., Klopper, W., and Helgaker, T. Electron correlation: The many-body problem at the heart of chemistry. *J. Comput. Chem.* 2007, 28, 1307–1320.

159. Schmidt, M.W. and Gordon, M.S. The construction and interpretation of MCSCF wavefunctions. *Annu Rev. Phys. Chem.* 1998, 49, 233–266.

160. Roos, B.O. The complete active space self-consistent field method and its applications in electronic structure calculations. *Adv. Chem. Phys.* 1987, 69, 399–445.

161. Andersson, K., Malmqvist, P.A., Roos, B.O., Sadlej, A.J., and Wolinski, K. Secondorder perturbation theory with a CASSCF reference function. *J. Chem. Phys.* 1990, 94, 5483–5488.

162. Andersson, K., Malmqvist, P.-A., and Roos, B.O. Second-order perturbation theory with a complete active space self-consistent field reference function. *J. Chem. Phys* 1992, 96, 1218.

163. Josefsson, I., Kunnus, K., Schreck, S., Föhlisch, A., de Groot, F.M.F., Wernet, P., and Odelius, M. *Ab initio* calculations of X-ray spectra: Atomic multiplet and molecular orbital effects in a multi-configurational SCF approach to the L-edge spectra of transition metal complexes. *J. Phys. Chem. Lett.* 2012, 3, 3565–3570.

164. Wernet, P., Kunnus, K., Schreck, S., Quevedo, W., Kurian, R., Techert, S., de Groot, F.M.F., Odelius, M., and Föhlisch, A. Dissecting local atomic and intermolecular interactions of transition-metal ions in solution with selective X-ray spectroscopy. *J. Phys. Chem. Lett.* 2012, 3448–3453.

# 5 Biological Catalysts

*Serena DeBeer*

## CONTENTS

In order to understand the mechanism of biological catalysis, one requires detailed knowledge of the changes that occur in the electronic and geometric structure of a metalloenzyme active site. Photon-in photon-out X-ray spectroscopy provides a means to obtain higher-resolution information than conventional X-ray absorption spectroscopy. This chapter highlights the latest results on biological catalysts and small molecule reference complexes using both one- and two-dimensional X-ray emission spectroscopy.

## 5.1 INTRODUCTION

In nature, metalloenzymes are able to carry out remarkable chemical transformations with a selectivity and efficiency that is often difficult or even impossible to mimic in the laboratory. Driven by a desire to use nature as an inspiration for rational catalysis design, the field of bioinorganic chemistry has had a long history of trying to understand the mechanism of metalloenzyme mediated chemical transformations. This means that one wants to understand how the metalloenzyme, in particular, the metal "active site" changes before, during, and after the chemical reaction. It is here that X-ray spectroscopy has played a pivotal role. Some of the earliest insights into the active site structures of metalloproteins came from early X-ray absorption spectroscopy studies (XAS), particularly in the extended X-ray absorption fine structure region (EXAFS). These studies provided some of the first hints of the complexity of the active site in nitrogenase (which is responsible for the reduction of $N_2$ to $NH_3$), as well as providing the first measure of Mn-Mn distances within the oxygen-evolving complex of Photosystem II.[1,2] In addition, EXAFS studies have provided insight into numerous enzymatic intermediates—including the active sites of laccase,[3] ribonucleotide reductase,[4] and methane monooxygenase,[5] to name only a few.

In recent years, the development of intense synchrotron beam lines equipped with high-resolution crystal analyzers, has opened up the application of photon-in photon-out X-ray spectroscopy to the field of bioinorganic chemistry.[6,7] Such setups enable the measurement of high-resolution X-ray emission data, allowing for both one-dimensional nonresonant X-ray emission (XES) and two-dimensional resonant X-ray emission (RXES) measurements to be performed. In this chapter, we first briefly review the methods and then summarize the recent applications of photon-in photon-out spectroscopy to biological systems and related small molecule model complexes. In our view, this is an area with tremendous potential for growth, and the advent of free electron lasers has opened up even greater possibilities for understanding the time-dependent transformations of metalloenzyme active sites.[8] Recent applications of dispersive X-ray emission to biological systems are highlighted and future possibilities are briefly discussed.

## 5.2  METHODS

Following the ionization of a core electron, electrons from higher levels will decay to refill the core hole. X-ray emission spectroscopy involves measuring all of the fluorescent photons, which result following the decay process. After the ionization of a $1s$ core electron on a transition metal absorber, the most likely event is that a $2p$ electron will decay to refill the $1s$ hole. This produces a so-called K$\alpha$ emission line, which will split into two lines due to $2p$ core hole spin orbit coupling. The next most likely event is that a $3p$ to $1s$ transition occurs, the so-called K$\beta$ emission line (Figure 5.1). While the $3p$ spin orbit coupling contributions will be much smaller than the $2p$ spin orbit coupling, additional contributions will arise due to $3p$-$3d$ exchange interactions for transition metals with unpaired $d$-electrons.[7] As a result, the K$\beta$ emission lines are sensitive to metal $d$-count and spin state. Finally, there is a very low probability (~1000 times less likely than the K$\alpha$ emission) that a valence electron will refill the $1s$ core hole (Figure 5.1). This gives rise to the K$\beta_{2,5}$ (ligand np to metal $1s$) and K$\beta''$ (ligand ns to metal $1s$) transitions, which are collectively referred to as the *valence-to-core region* or V2C XES. V2C XES is a sensitive probe of ligand identity, ionization potential, and protonation state.[9,10] In addition, it is a sensitive probe of metal-ligand bond lengths.[9,11] It thus provides a means to overcome many of the limitations of standard XAS measurements, as will be highlighted later in this chapter.

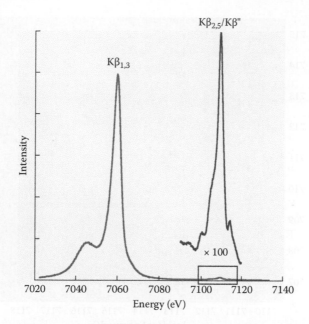

**FIGURE 5.1 (See Color Insert.)** The $K\beta_{1,3}$ and valence-to-core (iV2C or $K\beta_{2,5}$ + $K\beta''$) regions of an XES spectrum. (Reprinted from Lee, N. et al., 2010, *Journal of the American Chemical Society* 132, 9715.[9] With permission. Copyright 2010 American Chemical Society.)

To further enhance the selectivity of XES spectroscopy, one can also conduct a two-dimensional resonant measurement, referred to as *resonant X-ray emission* (RXES) or *resonant inelastic X-ray scattering* (RIXS).[7,12] Herein, we will use the term RXES, though either term may be encountered in the literature. An RXES measurement involves scanning through an X-ray absorption edge while simultaneously monitoring the resulting emission processes. Thus, for a *1s2p* RXES measurement with initial state of $1s^2 2p^6 3d^n$, one will access a $1s^1 2p^6 3d^{n+1}$ intermediate state and a $1s^2 2p^5 3d^{n+1}$ final state (Figure 5.2).

The difference between the initial and final state is the equivalent of a transition metal L-edge (i.e., a *2p* to *3d* transition) and thus has the ability to provide L-edge like information with a hard X-ray probe. This is of particular interest for biological solution chemistry, which is not amenable to in-vacuum measurements. A further benefit of *1s2p* RXES is that the use of a hard X-ray probe reduces the rate of radiation damage relative to soft X-ray measurement. We note that a cut through the RXES plane at constant emission energy also provides a means to obtain higher resolution XANES data.[13] This is often referred to as *high-energy resolution fluorescence detection* or HERFD XAS.[14]

Similarly, *1s3p* and *1s*-valence RXES measurements are possible. Such measurements provide the added possibility of spin and ligand-selectivity, respectively. However, *1s3p* RXES has received relatively less attention and to our knowledge the potential utility of *1s*-valence RXES has yet to be explored.

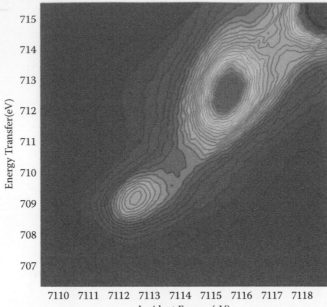

**FIGURE 5.2    (See Color Insert.)** A representative *1s2p* RXES plane. The *x*-axis corresponds to the incident energy and the *y*-axis to the energy transfer process, which yields an L-edge-like final state.

## 5.3    APPLICATIONS OF Kβ XES

Some of the earliest applications of Kβ XES to biological and biologically relevant systems were made by Cramer and coworkers.[15,16] The demonstration that Kβ XES measurements of proteins were feasible was first made in 1997 by Wang et al.[16] The authors showed that Kβ XES data could be obtained on the iron site of rubredoxin, as well as on the Mn site of Photosystem II, with concentrations as low as 40 ppm. While the study does not address the detailed interpretation of the spectra, it was essential in demonstrating the feasibility.

During the same time period, Cramer and coworkers also began conducting detailed model studies, which established important fingerprints for the application of Kβ XES to biological systems. In 1994, Peng et al. examined a series of Mn(II), Mn(III), and Mn(IV) model complexes and established the sensitivity of Mn Kβ XES to spin and oxidation state.[15] These studies formed a foundation for the direct application of Kβ XES to the oxygen-evolving complex of Photosystem II. In nature, a $Mn_4CaO_5$ metalloenzyme active site is responsible for the oxidation of water to form molecular dioxygen. The active site cycles through five S-states ($S_0$ to $S_4$), successively accumulating four oxidizing equivalents for the formation of $O_2$ (Figure 5.3).[17] One of the key questions in this field is whether the oxidizing equivalents are stored on the Mn ions or on the ligands. Hence, the sensitivity of Kβ XES to the number of unpaired electrons on the Mn ions is in this field a very useful tool.

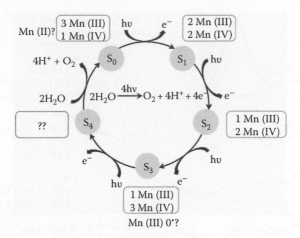

**FIGURE 5.3** **(See Color Insert.)** The S-State (or Kok cycle) in the oxygen-evolving complex of Photosystem II.

Using this approach, Messinger et al. were able to show evidence for a ligand-based oxidation in the $S_2$ to $S_3$ conversion.[18] By understanding where oxidizing equivalents are stored, the authors were able to discuss possible mechanistic implications. This study marked the start of many important applications of photon-in photon-out spectroscopy to the $Mn_4O_5Ca$ cluster of the OEC, as will be highlighted throughout this chapter.

## 5.4 APPLICATIONS OF V2C XES

As noted above, V2C XES provides a means to establish ligand identity (C, N, O), protonation state ($O^{2-}$, $OH^-$, or $H_2O$), and bond lengths. Notable applications of V2C XES include studies related to Photosystem II, nitrogenase, hydrogenase, and galactose oxidase. In addition, several detailed studies on related small molecule complexes have provided additional insights into the potential information content of V2C XES.

### 5.4.1 APPLICATIONS OF V2C XES RELEVANT TO PHOTOSYSTEM II

The work of Cramer, Glatzel, Bergmann, Yano, and Yachandra have provided powerful examples of the utility of photon-in photon-out spectroscopy for understanding complex systems, such as Photosystem II. In addition to the pioneering $K\beta$ XES mainline studies detailed in Section 5.3, Photosystem II is also the first biological system for which V2C XES was applied. In a 2010 study, Pushkar et al. demonstrated the presence of a weak but observable $K\beta''$ feature in the V2C XES spectrum of the S1-state of PSII (Figure 5.4).[19] Based on the comparison of these data to model complexes of known structure, they were able to assign this feature as a bridging mu-oxo group. As the presence of oxygenic groups and their protonation states is fundamental to understanding the mechanism of water oxidation, this study was an

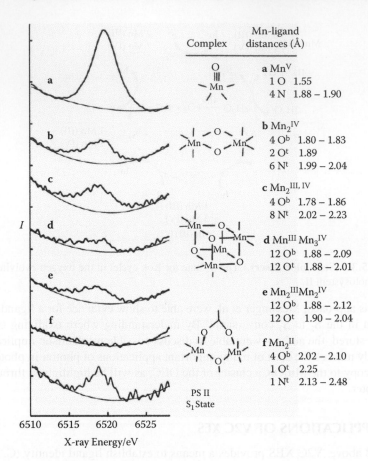

| Complex | Mn-ligand distances (Å) |
|---|---|
| O ‖‖‖ — Mn — / \ **a** | **a** Mn$^V$ <br> 1 O  1.55 <br> 4 N  1.88 – 1.90 |
| \ \|  / O \  \| / ⁻Mn⟨ ⟩Mn⁻ / \|  O \|  \ **b** | **b** Mn$_2^{IV}$ <br> 4 O$^b$  1.80 – 1.83 <br> 2 O$^t$  1.89 <br> 6 N$^t$  1.99 – 2.04 |
| **c** | **c** Mn$_2^{III, IV}$ <br> 4 O$^b$  1.78 – 1.86 <br> 8 N$^t$  2.02 – 2.23 |
| \| —Mn—O  \| \|  O ╀ Mn⟨ O ╀ Mn  \| —Mn— O \| **d** | **d** Mn$^{III}$ Mn$_3^{IV}$ <br> 12 O$^b$  1.88 – 2.09 <br> 12 O$^t$  1.88 – 2.01 |
| **e** | **e** Mn$_2^{III}$Mn$_2^{IV}$ <br> 12 O$^b$  1.88 – 2.12 <br> 12 O$^t$  1.90 – 2.04 |
| O   O \|\|   \|\| \ /  \ / Mn⟨   ⟩Mn \|  O  \| **f** | **f** Mn$_2^{II}$ <br> 4 O$^b$  2.02 – 2.10 <br> 1 O$^t$  2.25 <br> 1 N$^t$  2.13 – 2.48 |
| PS II S$_1$ State | |

6510    6515    6520    6525

X-ray Energy/eV

**FIGURE 5.4**  The Kβ″ XES data for a series of Mn complexes compared to the S1 state of Photosystem II. (Adapted from Pushkar, Y. et al., 2010, *Angewandte Chemie-International Edition* 49, 800.[19] With permission. Copyright 2010 Wiley-VCH Verlag GmbH & Co. KGaA, Weinheim.)

important demonstration of the utility of V2C XES. Another important outcome was the authors' observation that bridging carboxylate groups gave very weak or no Kβ″ features. This was attributed to the delocalization of the O 2s character over the carboxylate ligand and serves as an important observation of the limits of this method. We also note that the Kβ$_{2,5}$ data, that is, the higher energy region of the V2C XES spectrum, was not reported in this paper nor was a more quantitative assessment of these data given. However, the continued work by this team, as highlighted in the sections, which follow, gives promise that in due course V2C XES of the entire S-state cycle may become available.

In the interim, we note that further Mn model studies in the V2C XES region have demonstrated the utility of this method for determining protonation states.[20] A systematic study of Mn(IV) dimers in which the bridging ligands are varied from

$O^{2-}/O^{2-}$ to $O^{2-}/OH^-$ to $OH^-/OH^-$ was recently carried out by Lasselle et al. These studies show that a single protonation event results in a measurable shift in the $K\beta''$ feature, which can also be readily modeled within a one-electron density functional theory (DFT) picture. These results form a foundation for quantitative studies of the entire S-state cycle.

## 5.4.2 APPLICATIONS OF V2C XES RELEVANT TO NITROGENASE

Another notable application of V2C XES in biology is the recent studies by Lancaster et al. on the enzyme nitrogenase.[21,22] Nitrogenase affects the reduction of dinitrogen to ammonia using a 1Mo-7Fe-9S-C-containing cofactor, known as FeMoco (Figure 5.5). The presence of the $\mu_6$-carbon in the center of this cluster was recently revealed by V2C XES.[22] This finding is of particular note as the identity of the atom in the center of this cluster had eluded characterization by other methods for almost a decade. A 2002 crystal structure first revealed the presence of electron density at the center of the FeMoco cluster,[23] but whether it was a C, N, or O remained a mystery until 2011. Figure 5.5 shows the V2C XES data for the MoFe protein of FeMoco (which contains both the FeMoco cluster and an eight iron P-cluster), the data for isolated FeMoco in N-methyl formamide, and the data for the a delta-NifB gene deletion

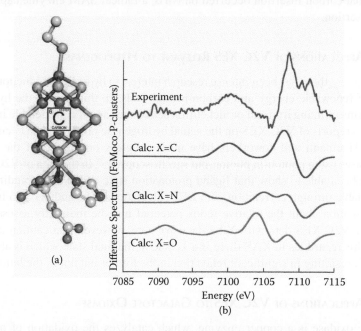

(a)

(b)

**FIGURE 5.5** (See Color Insert.) (a) The structure of the FeMoco cluster. (b) The experimental V2C XES difference spectra MoFe protein minus the p-cluster contribution. The calculations with a central C, N, or O are shown below. (Adapted from Lancaster, K.M. et al., 2011, *Science* 334, 974.[22] Copyright 2011 American Association for the Advancement of Science.)

mutant which contains only the P-cluster. These experimental data established that only when the FeMoco active site is present is there significant intensity at 7101 eV. By correlation of the protein data to model complexes of a known coordination environment, the $K\beta''$ feature was empirically assigned as a carbide. DFT calculations on the FeMoco with different central atoms (i.e., C, N, or O) were able to unambiguously assign the central atom as a carbon (Figure 5.5). This finding was also confirmed by a subsequent higher-resolution crystal structure and $^{13}$C-ESEEM data.[24]

The discovery that FeMoco had a carbon at its center built upon previous model studies of Delgado-Jaime et al.,[25] which focused on a six-iron cluster with a carbide at the center. The initial model studies correlated to calculations formed a solid foundation for assessing the sensitivity of V2C XES to changes in ligand identity. Deldgao-Jaime et al. also highlighted the strength of V2C XES relative to more traditional X-ray spectroscopic analyses, such as XANES and EXAFS approaches.

V2C XES has also provided insights into the assembly of the FeMoco cluster. In 2013, Lancaster et al. reported data on the NifEN precursor of nitrogenase.[21] This precursor corresponds to an earlier stage of assembly of the FeMoco cluster. Interestingly, these data showed that the carbon had already been inserted within the cluster at this stage of the biosynthesis. Based on this observation, it was possible to show that carbon insertion occurred on NifB, a radical SAM enzyme capable of methyl insertion.

### 5.4.3 Applications of V2C XES Relevant to Hydrogenase

In recent years, there has been intense research interest in hydrogen production in the context of renewable energy and carbon-neutral fuels. In this regard, the hydrogenase proteins utilizing iron and or nickel/iron active sites are of considerable interest. To date, no reports of V2C XES on the actual hydrogenase proteins have been made, however, Haumann and coworkers have made progress on models of the [FeFe] hydrogenases using photon-in photon-out spectroscopy.[26–28] In the area of V2C XES, they have been able to show that ligand protonation state and hydride binding may be identifiable through correlation with DFT calculations. The authors also indicate that information about the relative redox potential may be indirectly assessed, by combining V2C XES data with XAS data. We note, however, that caution must be taken in this regard, as in XAS there is a 1s hole in the final state which is absent in V2C XES, resulting in significant relaxation in the former but not in the latter.

### 5.4.4 Applications of V2C XES to Galactose Oxidase

Galactose oxidase is a copper enzyme, which catalyzes the oxidation of primary alcohols to aldehydes, coupled with the reduction of dioxygen to hydrogen peroxide. Although XES studies have not been reported on the enzyme system itself, studies of small molecule analogs indicate that the protonation of a phenolate ligand to copper results in subtle changes to V2C XES spectra.[29] For these model systems, the $K\beta''$ features are, in particular, weak likely due to delocalization of ligand 2s character over the entire phenolate ring resulting in reduced intensity.

## 5.4.5 INSIGHT INTO V2C XES FROM SMALL MOLECULE STUDIES

In Sections 5.4.1 through 5.4.4, examples where V2C XES was applied to biological systems or biomimetic systems have been discussed. However, advancing our understanding of these spectra requires detailed model studies. In this section, a few notable examples are briefly summarized.

In 2009, Glatzel and coworkers reported a V2C XES study on a series of molecular Mn model complexes.[10] They were able to show that the spectra could be modeled within a simple one-electron DFT framework, and also showed that the observed V2C XES transitions are dominated by metal p character mixed into ligand-based orbitals.

In 2010, Lee et al. conducted a systematic study on a series of Fe(II) and Fe(III) model complexes.[9] The authors used an approach similar to that of Glatzel and coworkers. In addition, they explored the indirect contribution of spin state to V2C XES spectra. As the metal ligand bond length increases, the metal p character mixed into the ligand orbitals decreases, resulting in lower intensity for high-spin complexes than for low-spin complexes. The authors also computationally explored the application of V2C XES to Compound II heme derivatives. The calculations indicate that a single protonation event at the heme active site should be detectable by V2C XES, though this remains to be experimentally demonstrated.

In 2011, Pollock and DeBeer examined the sensitivity of V2C XES to the nature of the bound ligand ($\sigma$-donating, $\pi$-donating, or $\pi$-accepting).[30] The authors showed that the intensity mechanism for V2C XES derives primarily from ligands, which by symmetry can interact with the metal np orbitals, thus conferring dipole allowed intensity to the V2C XES spectra. This also indicates that the shorter the metal-ligand interaction, the greater the intensity of the V2C XES, confirming earlier observations by Bergmann and DeBeer.[9,11] Pollock et al.'s study shows that the V2C XES provides a "map" of the ligand molecular orbitals that interact with the metal p-orbitals. As such, the authors were able to demonstrate that two possible proposed reaction pathways for the nonheme iron extradiol dioxygenase family of enzymes should be distinguishable by V2C XES. While the V2C XES spectra are only theoretical, they provide a very nice illustration of the potential power of this method for understanding biological catalysis.

In 2013, Pollock et al. further showed that V2C XES can be used as a probe of small molecule bond activation.[31] Using a series of iron model complexes in which the N-N bond varies from a formal triple bond to a fully cleaved bis-nitride species, the authors demonstrated that the experimental V2C XES energies can be directly correlated to the N-N bond strength (Figure 5.6). Specifically, an experimental peak, which corresponds to the N-N 2s-2s sigma antibonding molecular orbital, shows ~2 eV shift to lower energy upon N-N bond cleavage. The correlation between V2C XES peak energies and N-N vibrational frequencies was shown to be linear. This study indicates that V2C XES may be of utility as a time-resolved probe of small molecule activation.

Finally, we note that while all the examples discussed in this section have focused on 3d metals, applications of V2C XES to biologically relevant molybdenum dithiolene complexes have also been made by Bergmann, George, and Cramer.[32] Again,

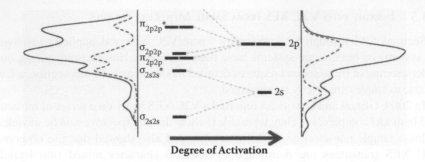

**Degree of Activation**

**FIGURE 5.6    (See Color Insert.)** The contribution of an $N_2$ derived ligand to the Fe V2C XES spectrum. The plot on the left corresponds to a short N-N bond. The plot on the right corresponds to a cleaved bis-nitride species. (Reprinted from Pollock, C.J. et al., 2013, *Journal of the American Chemical Society* 135, 11803.[31] With permission. Copyright 2013 American Chemical Society.)

the applications to the actual enzymatic systems remain to be made, however, the model studies clearly establish the potential power of this approach.

## 5.5    APPLICATIONS OF RXES

### 5.5.1    Applications of RXES to Photosystem II and Related Models

The team of Glatzel, Bergmann, Yano, and Yachandra have made significant contributions to the understanding and application of photon-in photon-out spectroscopy using Mn model complexes and the $Mn_4O_5Ca$ cluster of Photosystem II. In 2005, together with Cramer, they reported *1s2p* RXES on a series of Mn oxides and molecular $Mn(acac)_2(H_2O)_2$.[33] The data were interpreted within a multiplet-based approach and showed that the analysis of K-edge data could be greatly facilitated by obtaining higher-resolution RXES data. This approach has since been extended to the S0 to S3 states of Photosystem II.[34] The *1s2p* RXES spectral changes were shown to be very subtle, suggesting a strongly delocalized picture, where the ligands may also be involved in redox. The authors suggest that this delocalization may play a crucial role in O-O bond formation at the S4 state.

Another very interesting application to Photosystem II is the use of high-resolution Kα detection to obtain "range-extended" EXAFS.[35] While not formally an "RXES" or "RIXS" measurement in the conventional sense, this experiment involves using the same crystal analyzer setup to set an ~1 eV window on the Kα emission line. This experiment is thus an extension of HERFD XAS into the EXAFS region. It is of particular interest for Mn EXAFS on the photosystem, as these data are traditionally limited to ~11.5 Å⁻¹ in k-space due to the presence of Fe. In a 2005 study, Yano et al. demonstrated that EXAFS on Photosystem II could be obtained to k of 15.5 Å⁻¹, improving the resolution of the data from ~0.14 Å to ~0.09 Å.[35] This allowed for the separation of 2.7 and 2.8 Å Mn-Mn vectors, which previously was not possible. These studies were further extended in 2007 to oriented membranes of Photosystem II,[36] allowing for polarized range extended EXAFS data to be obtained.

## 5.5.2 Applications of RXES to Hydrogenase

Haumann and coworkers have applied *1s2p* RXES to FeFe models of hydrogenase.[26–28] By taking a vertical slice of a *1s2p* RXES they are able to obtain data on an L-edge-like final state. These data are then used to examine the changes in the unoccupied metal *d*-based orbitals in the presence and absence of a hydride. These data were used in conjunction with the V2C XES data (discussed in Section 5.4.3 above) to obtain potential insights into the HOMO-LUMO gap. More recently, Haumann and coworkers have also reported *1s3p* RXES.[28] In all cases, their data are correlated to calculations within a one-electron DFT or time-dependent DFT approach. We note that such a simplified approach does not fully capture resonant XES processes. It is clear, however, that this is an area where further theoretical advancements could greatly assist in the interpretation of experimental data.

## 5.5.3 Applications of Kβ-Detected XANES

Recently, Weckhuysen and Mijovilovich utilized Kβ detected XANES (i.e., a slice through a *1s3p* RXES plane) to obtain high-resolution XANES data Fe(II) and Fe(III) model complexes with a two-histidine-one-carboxylate binding motif.[37] Such a motif is found in numerous nonheme iron proteins. They are able to show that the high-resolution data aids in the quantitative interpretation of XANES data. The data were modeled using multiplet- and multiple-scattering-based approaches. The authors are able to show that the high-resolution data provides insight into the orientation of the carboxylate ligand.

Another interesting application of Kβ XANES involved the characterization of a mu-nitrido phthalocyanine complex $[PcFe^{+3.5}NFe^{+3.5}Pc]^0$, which is capable of oxidizing methane to methanol under ambient conditions.[38] It thus serves as a mimic of the enzyme methane monooxygenase. By examining the electronic structure difference between $[PcFe^{+3.5}NFe^{+3.5}Pc]^0$, $[PcFe^{+4}NFe^{+4}Pc]^+$, and $[PcFe^{+4}NFe^{+4}(Pc^{·+})]^{2+}$ the authors are able to show that the species are one-electron species in high-spin while the neutral and two-electron species are high-spin. They thus propose that in order to avoid undergoing a spin state change, the catalyst is more likely to do two-electron chemistry.

## 5.6 APPLICATIONS OF DISPERSIVE XES

In the preceding sections, all of the experiments were carried out using a crystal analyzer aligned on a Rowland circle in which the data were obtained in an energy-scanning mode. As discussed in Chapter 3, however, dispersive setups are also possible and allow for the entire XES spectrum to be obtained at a single shot.[39,40] This is thus of great interest for following kinetic processes. It is also a significant advantage for samples that undergo radiation damage. Recently, Davis et al. have shown that a dispersive setup with a flat crystal and position-sensitive detector can be used to obtain data from Photosystem II on the millisecond time scale at a 3rd-generation synchrotron source.[41] This setup has also been used to obtain a kinetic model for the rate of damage.[42]

Finally, very recently Kern et al. have reported the first femtosecond XES of Photosystem II obtained using a free electron laser.[8] By combining simultaneous X-ray diffraction and Kβ XES, the authors are able to show that their samples are intact during the free electron measurement. This is of significant importance as the ability to obtain data from undamaged biological samples using a free electron laser has been a subject of considerable debate. We note that at this point only Kβ mainline data have been reported, and it remains to be seen whether or not sufficient signal to noise can be achieved to observe weak V2C XES features. Regardless, it appears that advances in dispersive setups, as well as free electron laser developments, are poised to have a significant impact on our understanding of biological catalysis.

## 5.7 SUMMARY AND OUTLOOK

In this chapter, recent applications of photon-in photon-out spectroscopy to biological systems and related model complexes have been briefly summarized. It is clear that the ability to measure nonresonant and resonant emission has provided greater insights into the geometric and electronic structure of metalloprotein active sites, with notable progress in the areas of biological water oxidation, nitrogen reduction, and hydrogen production. In many cases, only models for the protein systems have been studied, and applications to the actual enzyme remain to be seen. This most likely reflects the application of high-resolution X-ray methods in biology, which are relatively new, and in the coming years, with the development of more dedicated beam lines for these measurements, one may expect photon-in photon-out spectroscopy to have an even broader impact on our understanding of biological catalysis.

## REFERENCES

1. Cramer, S.P., Hodgson, K.O., Gillum, W.O., and Mortenson, L.E. Molybdenum site of nitrogenase—Preliminary structural evidence from X-ray absorption spectroscopy. *Journal of the American Chemical Society* 1978, 100, 3398.
2. Kirby, J.A., Robertson, A.S., Smith, J.P., Thompson, A.C., Cooper, S.R., and Klein, M.P. State of manganese in the photosynthetic apparatus. 1. Extended X-ray absorption fine-structure studies on chloroplasts and di-mu-oxo bridged dimanganese model compounds. *Journal of the American Chemical Society* 1981, 103, 5529.
3. Lee, S.K., DeBeer-George, S., Antholine, W.E., Hedman, B., Hodgson, K.O., and Solomon, E.I. Nature of the intermediate formed in the reduction of $O_2$ to $H_2O$ at the trinuclear copper cluster active site in native laccase. *Journal of the American Chemical Society* 2002, 124, 6180.
4. Younker, J.M., Krest, C.M., Jiang, W., Krebs, C., Bollinger, J.M., and Green, M.T. Structural analysis of the Mn(IV)/Fe(III) cofactor of *Chlamydia trachomatis* ribonucleotide reductase by extended X-ray absorption fine structure spectroscopy and density functional theory calculations. *Journal of the American Chemical Society* 2008, 130, 15022.
5. Shu, L.J., Nesheim, J.C., Kauffmann, K., Munck, E., Lipscomb, J.D., and Que, L. An (Fe2O2)-O-IV diamond core structure for the key intermediate Q of methane monooxygenase. *Science* 1997, 275, 515.

6. Beckwith, M.A., Roemelt, M., Collomb, M.-N., DuBoc, C., Weng, T.-C., Bergmann, U., Glatzel, P., Neese, F., and DeBeer, S. Manganese K beta X-ray emission spectroscopy as a probe of metal-ligand interactions. *Inorganic Chemistry* 2011, 50, 8397.

7. Glatzel, P. and Bergmann, U. High resolution 1s core hole X-ray spectroscopy in 3d transition metal complexes—Electronic and structural information. *Coordination Chemistry Reviews* 2005, 249, 65.

8. Kern, J., Alonso-Mori, R., Tran, R., Hattne, J., Gildea, R.J., Echols, N., Gloeckner, C., Hellmich, J., Laksmono, H., Sierra, R.G., Lassalle-Kaiser, B., Koroidov, S., Lampe, A., Han, G., Gul, S., DiFiore, D., Milathianaki, D., Fry, A.R., Miahnahri, A., Schafer, D.W., Messerschmidt, M., Seibert, M.M., Koglin, J.E., Sokaras, D., Weng, T.-C., Sellberg, J., Latimer, M.J., Grosse-Kunstleve, R.W., Zwart, P.H., White, W.E., Glatzel, P., Adams, P.D., Bogan, M.J., Williams, G.J., Boutet, S., Messinger, J., Zouni, A., Sauter, N.K., Yachandra, V.K., Bergmann, U., and Yano, J. Simultaneous cemtosecond X-ray spectroscopy and diffraction of photosystem II at room temperature. *Science* 2013, 340, 491.

9. Lee, N., Petrenko, T., Bergmann, U., Neese, F., and DeBeer, S. Probing valence orbital composition with iron K beta X-ray emission spectroscopy. *Journal of the American Chemical Society* 2010, 132, 9715.

10. Smolentsev, G., Soldatov, A.V., Messinger, J., Merz, K., Weyhermueller, T., Bergmann, U., Pushkar, Y., Yano, J., Yachandra, V.K., and Glatzel, P. X-ray emission spectroscopy to study ligand valence orbitals in Mn coordination complexes. *Journal of the American Chemical Society* 2009, 131, 13161.

11. Bergmann, U., Horne, C.R., Collins, T.J., Workman, J.M., and Cramer, S.P. Chemical dependence of interatomic X-ray transition energies and intensities—A study of Mn K-beta" and K-beta(2,5) spectra. *Chemical Physics Letters* 1999, 302, 119.

12. Bergmann, U., Glatzel, P. X-ray emission spectroscopy. *Photosynthesis Research* 2009, 102, 255.

13. Hämäläinen, K., Siddons, D.P., Hastings, J.B., and Berman, L.E. Elimination of the inner-shell lifetime broadening in X-ray absorption spectroscopy. *Physical Review Letters* 1991, 67, 2850.

14. Link, P., Glatzel, P., Kvashnina, K., Smith, R.I., and Ruschewitzt, U. Yb valence states in YbC2: A HERFD-XANES spectroscopic investigation. *Inorganic Chemistry* 2011, 50, 5587.

15. Peng, G., Degroot, F.M.F., Hämäläinen, K., Moore, J.A., Wang, X., Grush, M.M., Hastings, J.B., Siddons, D.P., Armstrong, W.H., Mullins, O.C., and Cramer, S.P. High-resolution manganese X-ray fluorescence spectroscopy—Oxidation state and spin state sensitivity. *Journal of the American Chemical Society* 1994, 116, 2914.

16. Wang, X., Grush, M.M., Froeschner, A.G., and Cramer, S.P. High-resolution X-ray fluorescence and excitation spectroscopy of metalloproteins. *Journal of Synchrotron Radiation* 1997, 4, 236.

17. Yachandra, V.K., Sauer, K., and Klein, M.P. Manganese cluster in photosynthesis: Where plants oxidize water to dioxygen. *Chemical Reviews* 1996, 96, 2927.

18. Messinger, J., Robblee, J.H., Bergmann, U., Fernandez, C., Glatzel, P., Visser, H., Cinco, R.M., McFarlane, K.L., Bellacchio, E., Pizarro, S.A., Cramer, S.P., Sauer, K., Klein, M.P., and Yachandra, V.K. Absence of Mn-centered oxidation in the S-2 -> S-3 transition: Implications for the mechanism of photosynthetic water oxidation. *Journal of the American Chemical Society* 2001, 123, 7804.

19. Pushkar, Y., Long, X., Glatzel, P., Brudvig, G.W., Dismukes, G.C., Collins, T.J., Yachandra, V.K., Yano, J., and Bergmann, U. Direct detection of oxygen ligation to the Mn4Ca cluster of photosystem II by X-ray emission spectroscopy. *Angewandte Chemie-International Edition* 2010, 49, 800.

20. Lassalle-Kaiser, B., Boron, T., Krewald, V., Kern, J., Beckwith, M., Delgado-Jaime, M., Schroeder, H., Alonso-Mori, R., Nordlund, D., Weng, T.-C., Sokaras, D., Neese, F., Bergmann, U., Yachandra, V., DeBeer, S., Pecoraro, V., and Yano, J. Experimental and computational X-ray emission spectroscopy as a direct probe of protonation states in oxo-bridged Mn(IV) dimers relevant to redox-active metalloproteins. *Inorg. Chem.*, 2013, 52(22), 12915–12922.

21. Lancaster, K.M., Hu, Y.L., Bergmann, U., Ribbe, M.W., and DeBeer, S. *Journal of the American Chemical Society* 2013, 135, 610.

22. Lancaster, K.M., Roemelt, M., Ettenhuber, P., Hu, Y., Ribbe, M.W., Neese, F., Bergmann, U., and DeBeer, S. X-ray emission spectroscopy evidences a central carbon in the nitrogenase iron-molybdenum cofactor. *Science* 2011, 334, 974.

23. Einsle, O., Tezcan, F.A., Andrade, S.L.A., Schmid, B., Yoshida, M., Howard, J.B., and Rees, D.C. Nitrogenase MoFe-protein at 1.16 angstrom resolution: A central ligand in the FeMo-cofactor. *Science* 2002, 297, 1696.

24. Spatzal, T., Aksoyoglu, M., Zhang, L., Andrade, S.L.A., Schleicher, E., Weber, S., Rees, D.C., and Einsle, O. Evidence for interstitial carbon in nitrogenase FeMo cofactor. *Science* 2011, 334, 940.

25. Delgado-Jaime, M.U., Dible, B.R., Chiang, K.P., Brennessel, W.W., Bergmann, U., Holland, P.L., and DeBeer, S. Identification of a single light atom within a multinuclear metal cluster using valence-to-core X-ray emission spectroscopy. *Inorganic Chemistry* 2011, 50, 10709.

26. Leidel, N., Chernev, P., Havelius, K.G.V., Ezzaher, S., Ott, S., and Haumann, M. Site-selective X-ray spectroscopy on an asymmetric model complex of the FeFe hydrogenase active site. *Inorganic Chemistry* 2012, 51, 4546.

27. Leidel, N., Chernev, P., Havelius, K.G.V., Schwartz, L., Ott, S., and Haumann, M. Electronic structure of an FeFe hydrogenase model complex in solution revealed by X-ray absorption spectroscopy using narrow-band emission detection. *Journal of the American Chemical Society* 2012, 134, 14142.

28. Leidel, N., Hsieh, C.-H., Chernev, P., Sigfridsson, K.G.V., Darensbourg, M.Y., and Haumann, M. Bridging-hydride influence on the electronic structure of an FeFe hydrogenase active-site model complex revealed by XAES-DFT. *Dalton Transactions* 2013, 42, 7539.

29. Mijovilovich, A., Hamman, S., Thomas, F., de Groot, F.M.F., and Weckhuysen, B.M. Protonation of the oxygen axial ligand in galactose oxidase model compounds as seen with high resolution X-ray emission experiments and FEFF simulations. *Physical Chemistry Chemical Physics* 2011, 13, 5600.

30. Pollock, C.J. and DeBeer, S. Valence-to-core X-ray emission spectroscopy: A sensitive probe of the nature of a bound ligand. *Journal of the American Chemical Society* 2011, 133, 5594.

31. Pollock, C.J., Grubel, K., Holland, P.L., and DeBeer, S. Experimentally quantifying small-molecule bond activation using valence-to-core X-ray emission epectroscopy. *Journal of the American Chemical Society* 2013, 135, 11803.

32. Doonan, C.J., Zhang, L.M., Young, C.G., George, S.J., Deb, A., Bergmann, U., George, G.N., and Cramer, S.P. High-resolution X-ray emission spectroscopy of molybdenum compounds. *Inorganic Chemistry* 2005, 44, 2579.

33. Glatzel, P., Yano, J., Bergmann, U., Visser, H., Robblee, J.H., Gu, W.W., de Groot, F.M.F., Cramer, S.P., and Yachandra, V.K. Resonant inelastic X-ray scattering (RIXS) spectroscopy at the MnK absorption pre-edge—A direct probe of the 3d orbitals. *Journal of Physics and Chemistry of Solids* 2005, 66, 2163.

34. Glatzel, P., Schroeder, H., Pushkar, Y., Boron, T., III, Mukherjee, S., Christou, G., Pecoraro, V.L., Messinger, J., Yachandra, V.K., Bergmann, U., and Yano, J. Electronic structural changes of Mn in the oxygen-evolving complex of photosystem II during the catalytic cycle. *Inorganic Chemistry* 2013, 52, 5642.

35. Yano, J., Pushkar, Y., Glatzel, P., Lewis, A., Sauer, K., Messinger, J., Bergmann, U., and Yachandra, V. High-resolution Mn EXAFS of the oxygen-evolving complex in photosystem II: Structural implications for the Mn4Ca cluster. *Journal of the American Chemical Society* 2005, 127, 14974.

36. Pushkar, Y., Yano, J., Boussac, A., Bergmann, U., Sauer, K., and Yachandra, V. 2007. Structural changes in the oxygen evolving complex of photosystem II upon S-2 to S-3 transition. Polarized range-extended X-ray absorption spectroscopy of oriented photosystem II membranes in the S-1 state. *Photosynthesis Research* 2007, 91, 177.

37. Mijovilovich, A., Hayashi, H., Kawamura, N., Osawa, H., Bruijnincx, P.C.A., Gebbink, R.J.M.K., de Groot, F.M.F., and Weckhuysen, B.M. K as detected high-resolution XANES of FeII and FeIII models of the 2-His-1-carboxylate motif: Analysis of the carboxylate binding mode. *European Journal of Inorganic Chemistry* 2012, 1589.

38. Kudrik, E.V., Safonova, O., Glatzel, P., Swarbrick, J.C., Alvarez, L.X., Sorokin, A.B., and Afanasiev, P. Study of N-bridged diiron phthalocyanine relevant to methane oxidation: Insight into oxidation and spin states from high resolution 1s core hole X-ray spectroscopy. *Applied Catalysis B-Environmental* 2012, 113, 43.

39. Szlachetko, J., Nachtegaal, M., de Boni, E., Willimann, M., Safonova, O., Sá, J., Smolentsev, G., Szlachetko, M., van Bokhoven, J.A., Dousse, J.C., Hoszowska, J., Kayser, Y., Jagodzinski, P., Bergamaschi, A., Schmitt, B., David, C., and Luecke, A. A von Hamos X-ray spectrometer based on a segmented-type diffraction crystal for single-shot X-ray emission spectroscopy and time-resolved resonant inelastic X-ray scattering studies. *Review of Scientific Instruments* 2012, 83.

40. Alonso-Mori, R., Kern, J., Gildea, R.J., Sokaras, D., Weng, T.-C., Lassalle-Kaiser, B., Rosalie, T., Hattne, J., Laksmono, H., Hellmich, J., Gloeckner, C., Echols, N., Sierra, R.G., Schafer, D.W., Sellberg, J., Kenney, C., Herbst, R., Pines, J., Hart, P., Herrmann, S., Grosse-Kunstleve, R.W., Latimer, M.J., Fry, A.R., Messerschmidt, M.M., Miahnahri, A., Seibert, M.M., Zwart, P.H., White, W.E., Adams, P.D., Bogan, M.J., Boutet, S., Williams, G.J., Zouni, A., Messinger, J., Glatzel, P., Sauter, N.K., Yachandra, V.K., Yano, J., and Bergmann, U. Energy-dispersive X-ray emission spectroscopy using an X-ray free-electron laser in a shot-by-shot mode. *Proceedings of the National Academy of Sciences of the United States of America* 2012, 109, 19103.

41. Davis, K.M., Mattern, B.A., Pacold, J.I., Zakharova, T., Brewe, D., Kosheleva, I., Henning, R.W., Graber, T.J., Heald, S.M., Seidler, G.T., and Pushkar, Y. Fast detection allowing analysis of metalloprotein electronic structure by X-ray emission spectroscopy at room temperature. *Journal of Physical Chemistry Letters* 2012, 3, 1858.

42. Davis, K.M., Kosheleva, I., Henning, R.W., Seidler, G.T., and Pushkar, Y. Kinetic modeling of the X-ray-induced damage to a metalloprotein. *Journal of Physical Chemistry B* 2013, 117, 9161.

# 6 Heterogeneous Catalysts

## Jacinto Sá

## CONTENTS

The characterization of catalysts' electronic structure before, during, and after reaction is vital for the fundamental understanding of materials performance and stability. This chapter describes the latest results on heterogeneous catalysts' characterization using X-ray photon-in photon-out core level spectroscopy.

## 6.1 INTRODUCTION

This chapter focuses on the latest developments in X-ray photon-in photon-out core level spectroscopy applications for the understanding of heterogeneous catalysis. Catalytic performance is driven by the electronic structure of the valence shell. The

**169**

availability of valence orbitals to form chemical bonds, and thus taking part in the catalytic reaction depends on their electron occupancy and energy.[1] X-ray photon-in photon-out core level spectroscopy is a powerful tool to understand catalytic reactions because it enables us to map the entire electronic structure of the catalyst under reaction conditions. The chapter is divided into three sections, namely, characterization, adsorption, and reaction. The characterization section relates to studies in which the spectroscopy was used to determine the electronic structure of materials, and thus verify if the desired material was or was not successfully synthesized. The subsequent section (adsorption) reports advances in the use of X-ray photon-in photon-out core level spectroscopy to determine molecules adsorption geometry and strength. Substrate adsorption is the first step in any catalytic process, thus elucidation of its parameters is very valuable to understand catalytic performance especially when measurements are performed under real working conditions. The reaction section summarizes some of the X-ray photon-in photon-out core level spectroscopy performed on the entire catalytic cycle, which helped revealed aspects such as the active state of the catalyst, potential deactivation mechanisms, and so on.

## 6.2   CHARACTERIZATION

There are a multitude of studies using resonant inelastic X-ray scattering (RIXS) and its related techniques to unveil the electronic structure of catalysts. This section highlights some of the studies and recent developments to obtain the electronic states of catalytic active materials.

### 6.2.1   Transition Metal Oxides (3d Metals)

Transition metal oxides are an important class of catalysts, used in a variety of catalytic processes including photocatalysis. $1s$ X-ray absorption near edge structure (XANES) metal oxides spectra reveal features associated with the transition of the $1s$ core electron to the $4p$ conduction band (K-edges) and some small pre-edge features assigned to quadrupole transitions from the $1s$ core state to the empty $3d$-states.[2] The dipole selection rules state that the $1s$ core electron is excited to the lowest possible state with $p$-character, which in the case of the first-row of transition metal oxides is the $4p$-band composed of antibonding combinations of metal $4p$-states with the oxygen $2p$-states.

$1s2p$ RIXS pre-edge region measurements of $3d$ metal oxides yields information on spin state, oxidation state, electronic structure parameters such as the crystal field splitting and hybridization effects, and, indirectly, geometric information. For example, the $1s2p$ RIXS pre-edge region of CoO has three discernable features, which are associated to quadrupole $1s3d$ excitations from $3d^7$ ground state to the $1s^1 3d^8$ intermediate state, followed by the $1s2p$ X-ray emission decay into the $2p^5 3d^8$ final states.[3] However, there is no uniform interpretation of the pre-edge spectral features, that is, each metal has to be analyzed individually.[4]

Szlachetko and Sá[5] collected RIXS plane around Ti K-edge of $K\beta_{1,3}$ and valence-to-core transitions on a von Hamos spectrometer, effectively mapping the lowest

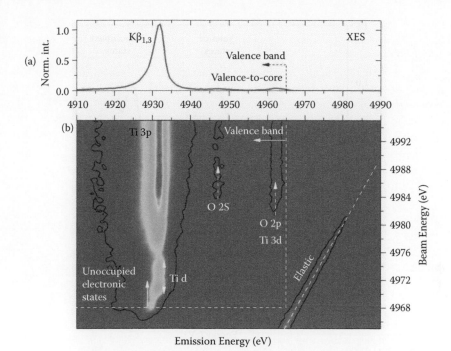

**FIGURE 6.1** **(See Color Insert.)** $TiO_2$ anatase RIXS map (b) and nonresonant XES spectrum (a). (Adapted from Szlachetko, J. and J. Sá, 2013, *Cryst. Eng. Comm.*, 15, 2583–2587.[5] With permission.)

unoccupied and highest occupied electronic states of $TiO_2$, respectively. In semiconductor terminology the lowest unoccupied electronic states relate to the conduction band and the highest occupied electronic states to the valence band, which were extracted from the RIXS plane based on HR-XAS and nonresonant XES spectra, respectively (Figure 6.1). The computed density-of-states (DOS) shows that the pre-edge structure (conduction band) is composed essentially of empty Ti $d$-states, whereas the occupied states just below the Fermi level have an equal contribution of O $p$- and Ti $d$-orbitals (Figure 6.2).

The authors also demonstrated that N-doping does not affect the conduction band states but it does dramatically change the valence states because substitutional N-doping leads to the hybridization of N and O $p$-orbitals, which shifts the valence band to higher energies, that is, band gap narrowing. The results show that RIXS can be used to determine semiconductors' band gap and electronic structure, which are often difficult to derive on doped materials due to interferences, such as color centers.

### 6.2.2 Transition Metal Oxides (*5d* Metals)

$5d$ transition metal catalysts play a crucial role in the chemical industry providing many products of importance for our daily life.[6] Platinum-based catalysts are indisputably the most important class of heterogeneous catalysts used by the chemical

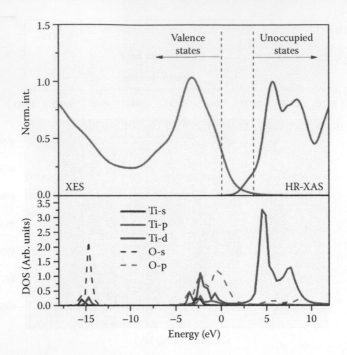

**FIGURE 6.2   (See Color Insert.)** Electronic structure of $TiO_2$ anatase extracted from experimental RIXS map (top) and computed DOS (bottom). (Adapted from Szlachetko, J. and J. Sá, 2013, *Cryst. Eng. Comm.*, 15, 2583–2587.[5] With permission.)

industry. They are used in a plethora of chemical processes for the production of fertilizers and plastics to pharmaceuticals.

It has been known since the late 1920s that the electronic structure plays a key role in the activity of heterogeneous catalysis,[7] thus its determination is vital toward establishing catalysts' reactivity.[8] One of the most effective ways to modify the electronic structure of $5d$ transition metals is to add a second metal.[9] A good example is the addition of $3d$ metals to Pt catalysts used in the oxygen reduction reaction (ORR) in polymer electrolyte membrane fuel cells. Pt L-edges probe $2p \rightarrow 5d$ transitions and the first intense spectral feature (whiteline) represents the unoccupied DOS in the $5d$-band.[10] Compared with $L_3$-edge ($2p_{3/2} \rightarrow 5d$), the $L_2$-edge ($2p_{1/2} \rightarrow 5d_{3/2}$) shows very small structures that are only detectable by high-energy resolution XAS.[11] The $L_2$-edge (ca. 13265 eV) is a higher energy transition, which enables determination of the electronic structure of Pt doped with $3d$ metals without interferences because the excitation energy is significantly different to the $3d$ metals, more than in the case of $L_3$-edge (ca. 11565 eV).

Anniyev et al.[12] used Pt $L_2$-edge measurements to establish the effect of Cu addition to Pt $d$-DOS. The high energy resolution fluorescence detection (HERFD) experiments were performed on Pt and $Cu_3Pt$ foils with a Johann-type spectrometer. The addition of Cu was found to decrease dramatically Pt whiteline intensity, synonymous of a reduction in the unoccupied $d$-DOS. The observation was rationalized on the basis of changes in $d$-$d$ hybridization, and therefore $d$-bandwidth due to the formation of an alloy with metals with different $d$-level binding energies. The $Cu_3Pt$ catalyst had better catalytic performance on ORR compared to the bare Pt. The authors argued that the downward shift of Pt projected $d$-DOS, measured with the Pt $L_2$-edge HERFD experiments, causes a weakening of the chemical bond between oxygen species and surface Pt, thus enhancing the ORR reactivity.

Gold catalysis is an emerging technology, which has been fueled by the discovery that finely dispersed gold nanoparticles are highly active, and have selectivities exclusive to it when compared with other $5d$ transition metals.[13] This was surprising since bulk gold is inert. In its metallic state, gold has a filled $5d$-band ($Xe4f^{14}5d^{10}6s^1$), so according to the simple $d$-band model it does not possess unoccupied $d$-states able to coordinate to reactants. The dramatic change in catalytic performance when the gold is present as a nanoparticle suggests a change in $5d$ occupancy.

A single metallic gold atom shows no Au $L_3$-edge transition since the $d$-band is full, however, when the atoms form particles, the $s$-, $p$-, and $d$-orbitals hybridize, which pushes some of the $d$-orbitals to levels above the Fermi level, thus changing the electron configuration from $5d^{10}6sp^1$ to $5d^{10-x}6sp^{1+x}$, and consequently the appearance a $L_3$-edge whiteline.[14] This was exploited by Van Bokhoven and Miller[15] to determine the $d$-DOS of Au as a function of particle size. They reported an increase of $d$-DOS of about 0.2 electrons for 1 nm particles with respect to bulk gold. Furthermore, the $d$-band in smaller particles was found to be narrower and shifted closer to the Fermi level, thus justifying the higher heats of adsorption and catalytic reactivity of smaller nanoparticles. In conclusion, the authors revealed that the number of atoms that contribute to the $d$-band affects the occupancy and energy of the Au $d$-band.

### 6.2.3 LANTHANIDES

Rare earth systems are a very important class of materials for a variety of fields, including catalysis,[16] fuel cells,[17] bioscience,[18] material science,[19] biology,[20] and mineralogy.[21] Most lanthanides are found in a trivalent valence state but some exist with a different oxidation state, for example, Ce, Yb, Eu, and Sm. Heavy fermions with these elements have a partially filled valence shell, which confers on them extraordinary properties.[22]

Ceria-based materials are catalytically active in exhaust pollutants abetment,[23] total or partial oxidation of hydrocarbons,[24] and elimination of CO from hydrogen stream (PROX reaction).[25] $CeO_2$ reactivity is commonly assigned to the low redox potential of the $Ce^{3+}/Ce^{4+}$ couple and to its high oxygen storage capacity,[26] which is believed to change with particle size. Paun et al.[27] performed Ce $L_3$-edge HERFD-XANES on $CeO_2$ nanoparticles with different particle sizes to evaluate the role of size in ceria in the stabilization of $Ce^{3+}$. A characteristic Ce $L_3$-edge HERFD-XANES spectrum of $CeO_2$ is depicted in Figure 6.3.

The main features in Ce $L_3$-edge are due to transitions from $2p_{3/2}$ to $5d_{5/2}$ (features $A_1$, $A_2$, B, and C), while the pre-edge around 5720 eV relates to the $2p$ to $4f$ process (feature D).[28] The screen (B and C) and unscreened ($A_1$ and $A_2$) edge features provides information on $4f$ population (oxidation state) and their doublets structure reveals the crystal field splitting of $5d$-orbitals. With respect to the influence of particle size in the Ce $L_3$-edge HERFD-XANES spectrum, Paun and coworkers'

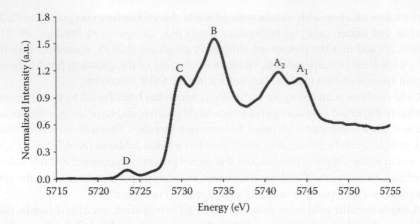

**FIGURE 6.3**   Ce $L_3$-edge HERFD-XANES spectrum of $CeO_2$ nanoparticles.

experiments revealed that under ambient conditions the average oxidation state is $Ce^{4+}$, and the amount of $Ce^{3+}$ is not influenced by particle size. They detected small differences on $Ce^{3+}$ concentration but that was related to the use of $Ce^{3+}$ precursors used in the synthesis. Finally, the authors reported a significant reduction of $CeO_2$ under intense X-ray irradiation, thus requiring the measurement to be carried out using a fast shutter to minimize beam damage.

## 6.3   ADSORPTION

Reactant adsorption is the initial step of every catalytic reaction. The catalytic output is intimately related to adsorption parameters, such as molecules adsorption geometry and strength, thus their manipulation can effectively change the catalytic yield.[29] Reactant adsorption is affected by the adsorption site's electronic and/or geometric structure, however, the former is the most effective in achieving fine tuning since bond formation between reactants and the active site is governed essentially by the catalyst electronic states, which control both molecules' adsorption strength and geometry.

Adsorption leads to the formation of new electronic states, composed by bonding and antibonding states. When the antibonding states are shifted to energy levels above the Fermi level, the molecule bonds to the metal site.[8] Hammer and Nørskov[30] proposed an adsorbate-metal bonding model based on $d$-band theory. The model suggests that adsorption energy depends on the energy position of the $d$-band with respect to the Fermi level. As a consequence, metals with the $d$-band closer to the Fermi level are more prone to adsorb molecules than those with the $d$-band farther from the Fermi level. Often, it is difficult to extract information about reactant's adsorption under realistic conditions especially for liquid phase reactions that are often carried out at high pressure and temperature. This section contains some examples where RIXS and its related techniques were used to determine a reactant's adsorption geometry and/or adsorption strength under realistic conditions.

## 6.3.1 GAS PHASE ADSORPTION

### 6.3.1.1 CO Adsorption on Pt Catalysts

Carbon monoxide oxidation is a very important process in exhaust catalytic converters.[31] Another important process is the preferential oxidation of CO in hydrogen reach streams (PROX reaction), which decreases CO content in $H_2$ feeds, thus minimizing catalyst poisoning, in particular hydrogen fuel cells anode poisoning. This is the most acute problem, and the prime reason stopping hydrogen fuel cells proliferation.[32] A plethora of catalysts are used to remove CO but invariably the best performing systems contain platinum. This justifies the large number of reports on CO adsorption on Pt, and part of the reason for Gerhard Ertl's 2007 Nobel Prize in chemistry.[33]

Safonova et al.[34] reported changes in the HERFD spectra due to CO adsorption on Pt measured at Pt $L_3$-edge ($2p_{3/2} \rightarrow 5d$). The Pt $L_3$-edge spectrum probes the unoccupied states of Pt $5d$-states, which is affected by molecular adsorption resulting from orbital rehybridization, charge transfer, metal-adsorbate scattering, and differences in metal-metal scattering. When performed in high-energy resolution mode, one is able to detect small variations in the DOS, making it an ideal technique to probe adsorption. Furthermore, the use of hard X-rays allows the experiments to be carried out under reacting conditions due to the probe high penetration. The $Pt/Al_2O_3$ Pt $L_3$-edge HERFD spectrum is dominated by the whiteline feature with a maximum at ~11567 eV. Upon CO adsorption the ionization threshold shifts to higher energy and the whiteline intensity increases. Furthermore, a doublet feature now characterizes the whiteline. Based on FEFF calculations,[35] the authors assigned the observed changes to the adsorption of CO on Pt at the atop position instead of bridged and/or hollow site. Orbital analysis revealed that in the case of Pt with CO on the atop position, the Pt $d$-orbitals overlap with the $2\pi^*$-orbitals of C and O atoms, forming an antibonding state above the Fermi level responsible for the double second feature. The transition has a significant metal-to-ligand charge transfer.

RIXS measurements (Figure 6.4) revealed a few electron volts energy transfer arising from valence-band excitations.[36] For bare Pt nanoparticles (without CO) the elastic peak and the valence band merged together, indicative of Pt in metallic state, with the Fermi level located within the partially filled $d$-band. CO adsorption leads to an increase in intensity above 4 eV energy transfer and broadening of the energy distribution. A gap between elastic peak and lowest unoccupied electronic states emerges. Upon CO adsorption on Pt, bonding, nonbonding, and antibonding orbitals are formed. There is a strong hybridization between Pt $d$-orbital and CO $p$- and $s$-orbitals.[37] Thus, the valence electrons populate deeper binding energies due to the formation of a bond between CO and Pt. In moving the $d$-band to lower energies, Pt becomes less prone to form bonds with other reactants, that is, surface poisoning, thus affecting catalytic performance.[38] The experiments were carried out with a Johann-type spectrometer on a Köningsberger *in situ* cell.[39]

Sá et al.[40] demonstrated that CO adsorption geometry on Pt could be modified by an external magnetic field, when Pt is supported on a carbon capped Co nanocore.

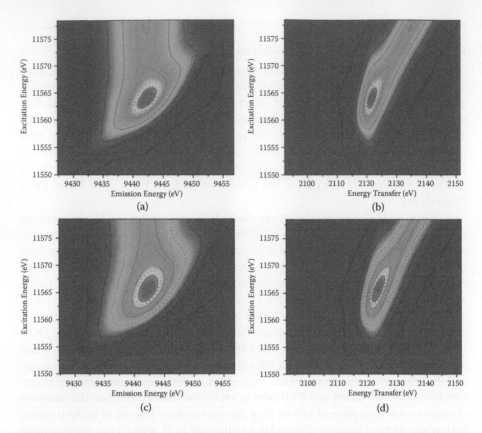

**FIGURE 6.4** **(See Color Insert.)** Calculated RIXS maps of Pt $L_3$-edge. (a) RIXS of bare $Pt_6$; (b) energy transfer RIXS of bare $Pt_6$; (c) RIXS of CO adsorbed atop on $Pt_6$; (d) energy transfer RIXS of CO adsorbed atop on $Pt_6$.

In the presence of a magnetic field part of the atop CO moved to the bridge position, which is a nonreactive state.[41] This represents a decrease in adsorption strength of roughly 0.1 eV,[42] enabling CO desorption at lower temperature. The conclusion was based on Pt $L_3$-edge RIXS measurements (Figure 6.5), using a von Hamos-type spectrometer with a capillary-type reactor.[43] In this case, conventional vibrational spectroscopy was not suitable due to the specimen's strong infrared absorption.

### 6.3.1.2   $H_2$ and $O_2$ Adsorption on Pt Catalysts

A large number of heterogeneous catalytic processes employ $H_2$ and/or $O_2$, either in pretreatments and/or as reagents, thus discrimination of changes induced in the electronic structure of a catalyst due to their adsorption is very important. Vicente et al.[44] followed the adsorption of $H_2$ and $O_2$ on Pt supported catalysts. Hydrogen adsorption led to a small increase in the whiteline intensity and blue shift (~ 0.5–0.8 eV). The magnitude of the changes depended on the catalyst support. Oxygen adsorption led to significant changes in both whiteline intensity and position. The whiteline intensity doubled with respect to the Pt reduced signal and shifted from 11567.3 eV to

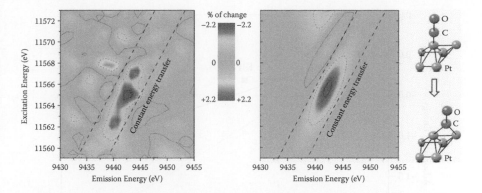

**FIGURE 6.5** (See Color Insert.) Pt $L_3$-edge $\Delta$-RIXS due to the presence of a 50 mT magnetic field on Pt on Co with adsorbed CO (Field OFF–Field ON). (Left) Experimental map differences measured *in situ*; (right) calculated map differences. (Reproduced from Sá et al., 2013, *Nanoscale* 5, 8462.[40] With permission.)

11568.4 eV (a 1.1 eV shift), suggesting partial oxidation of Pt. The extent of Pt oxidation depends on reaction parameters, such as temperature, metal particle size, and so forth. Nevertheless, the measurements show that the technique is sensitive to the adsorption of small molecules, such as $H_2$ and $O_2$, which are an integrant component of a multitude of catalytic processes.

### 6.3.1.3 N₂O Adsorption on Fe-ZSM-5

The combination of Fe-ZSM-5 and $N_2O$ is able to oxidize benzene with very high selectivity.[45] A pertinent question is how the $N_2O$ adsorbs on the Fe center, consequently determining the reactive and selective oxygen species involved in the process. Proposals for the active site include Fe(III)-O⁻ radical[46] or Fe(IV) = O ferryl species.[47] Pirngruber et al.[48] performed RIXS at Fe K-edge (*1s* → *3d* transition) and analyzed the *3p* → *1s* (K$\beta$) fluorescence line because the exchange interaction of the *3p-* with the *3d*-orbitals is stronger than of the *2p*-orbitals. The K$\beta$ emission was resolved into a main emission line (K$\beta_{1,3}$) and a satellite (K$\beta'$) at lower energy. The absence of the pre-edge peak in Fe K-edge XANES spectrum recorded on the K$\beta'$ emission line after adsorption of $N_2O$ confirms that the active site is the Fe(III)-O⁻ radical ($3d^5$), as suggested by Mössbauer spectroscopy.[49] The results suggest that the Fe-ZSM-5/$N_2O$ high activity is due to the formation of active oxygen species bound to Fe(III) and not to the high oxidation state of iron. The experiments were carried out with a Johann-type spectrometer using a 1 mm capillary-type reactor.

### 6.3.2 LIQUID PHASE ADSORPTION

Reactant and product adsorption–desorption characteristics are routinely evaluated with a variety of spectroscopic techniques, when the reaction is carried out in the gas phase. However, this is considerably harder to do in liquid phase, mainly due to the presence of solvent. The problem is further exacerbated if one considers that most of these reactions are performed in pressurized autoclaves.

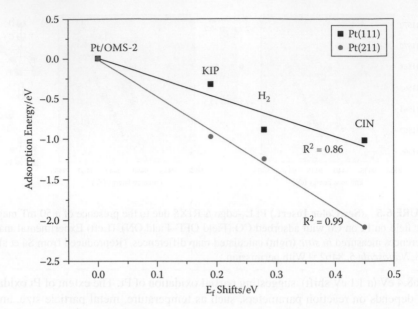

**FIGURE 6.6    (See Color Insert.)** Correlation between the experimentally measured Fermi level shifts and the DFT calculated adsorption energies for Pt(111) (black) and Pt(211) (red) surfaces. (From Manyar et al., 2013, *Catal. Sci. Technol.* 3, 1497–1500.[50] With permission.)

Manyar et al.[50] used HERFD-XANES to determine the changes in Pt electronic structure due to the adsorption of an α,β-unsaturated aldehyde or ketone. The interesting aspect of this study relates to the fact that the measurements were carried out under pressure (10 bar) and temperature (100°C) conditions at which the α,β-unsaturated aldehyde or ketone selective hydrogenation reactions take place.[51] The HERFD-XANES experiments were carried out in a homemade autoclave with a window and a PEEK (polyether ether ketone) insert.[52]

The adsorption ketoisophorone (KIP), $H_2$, and cinnamaldehyde (CIN) on Pt/OMS-2 caused a blue shift of the Fermi level ($E_f$) energy with respect to the reduced catalyst of 0.19 ± 0.09, 0.28 ± 0.08, and 0.45 ± 0.09 eV, respectively. The Fermi level shifts correlated remarkably well with the adsorption energies calculated via density functional theory (DFT), as one can see in Figure 6.6. The correlation is particularly clear on the Pt(211) surface, which describes better the Pt nanoparticles. The combination of HERFD-XANES experiments and DFT calculations provided a deeper understanding of catalytic reactivity. The Pt/OMS-2 selectively reduces the C=C in the case of ketoisophorone but hydrogenates the C=O in cinnamaldehyde.[51]

## 6.4    REACTION

A plethora of catalytic reactions have been followed by *in situ* with high-energy resolution XAS/XES. The criterion for this section is to report studies covering a broad range of catalytic processes and experimental conditions, which readers can associate with and hopefully adapt to study their own systems. The examples were chosen based on their novelty and adaptability to the study of other catalytic systems.

## 6.4.1 GAS PHASE PROCESSES

### 6.4.1.1 Propane Dehydrogenation

Dehydrogenation of light alkanes, such as propane, is of great interest due to the growing demand of propene as an important case chemical intermediate.[53] Pt-based catalysts are the most widely used system; however, these catalysts suffer from fast deactivation due to coke formation.[54] Sn is often added to alter the product distribution by inhibiting undesired reactions, such as hydrogenolysis, isomerization, cracking, and coke formation, thus enhancing catalyst selectivity and stability.[55] Despite being widely used as promoter, Sn promotion mechanism is not fully understood.

Iglesias-Juez et al.[56] performed an *in situ* HERFD-XANES study of Pt and Pt-Sn supported on $Al_2O_3$, aimed at understanding the effect of Sn in Pt electronic structure. The presence of a propane feed modified the $Pt/Al_2O_3$ $L_3$-edge difference spectra ($\Delta\mu$ = measured spectrum in propane spectrum of metal collected at same temperature) at 100°C, revealing a narrow negative peak at 11563 eV, and a broad positive peak at 11567 eV. The negative peak results from an edge shift, whereas the positive is related to an enhancement of the signal. The feature's intensity increased when temperature was increased to 140°C, after which all spectral differences decrease intensity, and the 11567 eV contribution shifted slightly to lower energy. The spectral changes observed with $Pt-Sn/Al_2O_3$ were significantly different to $Pt/Al_2O_3$. First, the negative peak observed in $Pt/Al_2O_3$ $\Delta\mu$ spectra (~11563 eV) was absent in the Sn-containing catalyst. In fact, in the case of the $Pt-Sn/Al_2O_3$ there was an increase of the intensity due to a broad absorption that overlaps with the 11567 eV. Furthermore, a very broad negative contribution appeared between 11570 and 11580 eV. Raising the temperature led to an increase of the positive features. After 120°C, the features diminished and shifted to lower energy. The observations clearly indicate that propane and Sn affect the electronic states of Pt.

The changes in Pt signal in the presence of propane were ascribed to either the presence of a π-(back) bonded (actually combined σ- and π-bonded) propene species and/or adsorbed hydrogen in Pt. The observation of doubly bonded (σ- and π-[back] bonded) suggested the formation of propyl species on Pt, which are difficult to remove (product poison). These species are known precursors for coke formation, which leads to a more permanent surface poisoning. The addition of Sn shifted the Pt edge to lower energy indicative of charge transfer from Pt to Sn (Pt becomes electron poor). This prevents π-(back) bonded interactions, that is, formation of σ-bonded Pt-propene complexes. The σ-bonded Pt-propene complexes are weaker than σ- and π-(back) bonded complexes, thus the interaction between product and substrate is weaker, which decreases the coke formation.

Coke formation is still a problem even when Sn is added to the catalyst formulation. A common practice to regenerate the catalyst is to switch the feed from propane to air followed by treatment in hydrogen. The air should burn out the carbon deposits

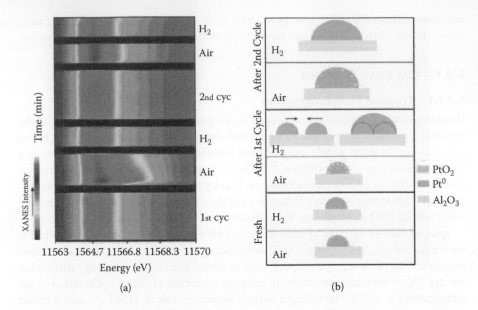

**FIGURE 6.7 (See Color Insert.)** (a) Intensity contour map of the HERFD-XANES of the Pt/Al$_2$O$_3$ catalyst acquired at 600°C during the first two propane dehydrogenation-regeneration cycles. (b) Schematic model of Pt species during the different treatment stages for the Pt/Al$_2$O$_3$ catalyst. (Reproduced from Iglesia-Juez et al., 2010, *J. Catal.* 276, 268–279.[56] With permission.)

and the hydrogen should reduce the metal. However, the regeneration cycles often change the catalyst and can be themselves a motive for catalyst deactivation.

In the same study, Iglesias-Juez et al.[56] study the effect of the regeneration cycles on the Pt electronic structure. The results are summarized in Figures 6.7 and 6.8 for Pt/Al$_2$O$_3$ and Pt-Sn/Al$_2$O$_3$, respectively. In the case of Pt/Al$_2$O$_3$, upon reduction of the calcined catalyst (fresh) Pt was present as metallic small particles. The first regeneration cycle led to the formation of partially oxidized small particles (under air), which sintered into larger particles as soon as the feed was changed to hydrogen. Subsequent regeneration cycles led to the formation of partially oxidized particles (in air) and their reduction (in hydrogen); however, the particle size did not change significantly. In the case of Pt-Sn/Al$_2$O$_3$, the reduction steps (in hydrogen) led to the formation of Pt-Sn alloy structures. This takes place due to the reduction of SnO$_2$ layers, which migrate from the support to the metal during the oxidation step (air). The regeneration cycles also sintered Pt particles; however, the effect was much less pronounced than on the unpromoted catalyst. Even after the 10th regeneration cycle the particles were still smaller than in the case of Pt/Al$_2$O$_3$, after the first regeneration cycle. This confirms tin dual function, namely, minimizing coke formation and preventing sintering.

### 6.4.1.2 Methane-to-Methanol Processes

Methane-to-methanol processes are of major importance for the chemical industry. Methanol is used in the production of several other industrially relevant chemicals

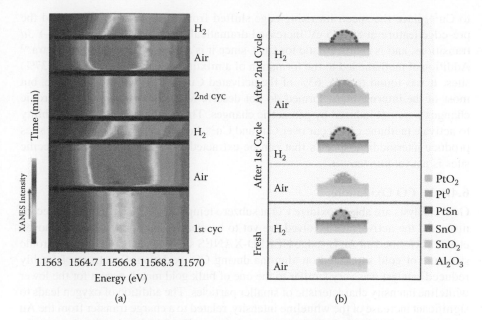

**FIGURE 6.8 (See Color Insert.)** (a) Intensity contour map of the HERFD-XANES of the Pt-Sn/Al₂O₃ catalyst acquired at 600°C during the first two propane dehydrogenation-regeneration cycles. (b) Schematic model of Pt species during the different treatment stages for the Pt-Sn/Al₂O₃ catalyst. (Reproduced from Iglesia-Juez et al., 2010, *J. Catal.* 276, 268–279.[56] With permission.)

and as fuel. The processes also convert a gas into a liquid, which is significantly easier and a cost-effective way of transporting methane.

The most common way to produce methanol from methane is via the process developed in 1966 by Imperial Chemical Industry, in which they convert syngas ($CO+H_2$), originated from partial oxidation of methane, into methanol at 250°C and high pressure 50–100 bar over a $Cu/ZnO/Al_2O_3$.[57] Despite catalyst longevity, there are numerous groups trying to establish a reaction mechanism and active site identity. Kleymenov et al.[58] performed a HERFD-XANES study of a commercial $Cu/ZnO/Al_2O_3$ under industrial feed composition, temperature (~250°C), and pressure (~6 bar). The study was performed in a capillary reactor and with a Johann-type spectrometer. Despite identifying the presence of $Cu^+$ during the reduction step (pretreatment), methanol synthesis only started when all the Cu sites were reduced to the metallic state. The Cu remained in the metallic phase throughout the reaction, however, activity decreased due to carbon deposition. The authors did not detect the formation of a Cu-Zn alloy, which was previously proposed as a possible active site.[59]

Methane can be partially oxidized to methanol under mild conditions over pre-activated Cu-exchanged zeolites.[60] However, the product is strongly adsorbed, thus methanol is obtained via extraction.[61] Alayon et al.[62] used HERFD-XANES to determine the electronic structure of pre-activated Cu-MOR (MOR = mordenite) during methane conversion. Catalyst activation involved treating the material in oxygen at 450°C. The pretreatment conditions lead to the oxidation of the metal site to $Cu^{2+}$, since the spectrum rising edge shifted from 8988 eV to 8986 eV, and the pre-edge feature at 8976.5 eV increased dramatically. Pre-edge relates to $1s \rightarrow 3d$ transitions, and is characteristic for $Cu^{2+}$, since it is absent in $Cu^+$ and $Cu^0$ spectra.[63] Addition of methane led to the formation of a mixture of $Cu^+$ (63%) and $Cu^{2+}$ (37%) sites. It was found that ca. 63% of the activated Cu-MOR react with methane, but most of the intermediates formed do not desorb. Most importantly, the electronic changes were not induced by geometric changes. The results suggest that the ability to activate methane can occur over $Cu^+$ and $Cu^{2+}$, however, only a few of these sites produce intermediate species that can be extracted. The identity of those specific sites is not yet known.

### 6.4.1.3   CO Oxidation

Gold catalysts are able to oxidize CO at subzero temperatures.[64] The reaction mechanism and the active sites involved are yet to be clearly identified.[65] Van Bokhoven et al.[66] performed an Au $L_3$-edge HERFD-XANES study to determine the electronic structure of gold supported on alumina during CO oxidation. Spectrum of freshly reduced catalyst was very similar to the one of bulk gold metal, except for the lower whiteline intensity characteristic of smaller particles. The addition of oxygen leads to significant increase of the whiteline intensity, related to a charge transfer from the Au $d$-band to the $2\pi^*$-orbital of oxygen, decreasing Au DOS and activating the oxygen, thus suggesting formation of a partially oxidized gold structure ($Au_yO_x$), accounting for roughly 15% of the Au in the catalyst. Subsequent addition of CO leads to a reduction of whiteline to a value close to the reduced catalysts but not exactly, suggestive that some of the gold is decorated with CO.[67] Based on the HERFD-XANES results, the authors proposed a two-step mechanism, according to the equations below:

$$Au^0 + O_2 \rightarrow Au_yO_x \tag{6.1}$$

$$Au_yO_x + CO \rightarrow Au^0 + CO_2 \tag{6.2}$$

### 6.4.2   LIQUID PHASE PROCESSES

### 6.4.2.1   Water Denitration

Water denitration relates to the catalytic removal of nitrates from drinking water. The occurrence of nitrate excess in water is due to the excessive use of fertilizers, and disposal of untreated waste effluents from certain industries. Nitrate pollution leads to eutrophication of water reservoirs, and serious human health problems, such as cancer, hypertension, and the commonly known blue-baby disease.[68] The research has been focused in the screening and/or improvement of bimetallic systems composed of a noble metal (Pt, Pd) and a transition metal (Cu, Sn), which decompose the nitrates into harmless nitrogen via a stepwise mechanism (Equation 6.3). The main drawback of the process is the formation of ammonium, which has even stricter limits in drinking water.

$$NO_3^-{}_{aq} \rightarrow NO_2^-{}_{aq} \rightarrow [NO]_{ads} \rightarrow N_2{}_{g} + NH_4^+{}_{aq} \tag{6.3}$$

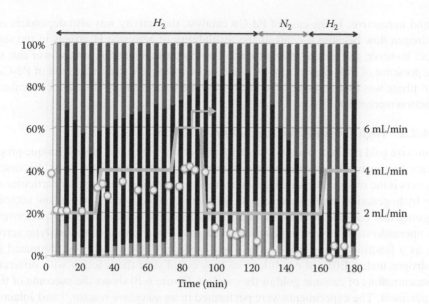

**FIGURE 6.9** **(See Color Insert.)** Deconvolved Cu K-edge HERFD-XAS spectra of Pt-Cu/ $Al_2O_3$ and nitrate conversion versus gas flow as a function of time stream. (Black) $Cu^0$, (red) CuO, (blue) Pt-Cu alloy; (yellow) gas flow; (') nitrate conversion. (Adapted from Sá, J. et al., 2012, *Catal. Sci. Technol.* 2, 794–799.[69] With permission.)

Sá et al.[69] performed an operando study to evaluate the electronic structure of the transition metal (Cu) as a function of reaction conditions. They used a capillary-type reactor and Johann-type spectrometer. The results for Pt-Cu bimetallic catalyst are summarized in Figure 6.9. The HERFD-XANES is plotted against nitrate conversion, in order to establish a relation between structure and activity. It should be mentioned that calibration with standard samples showed that HERFD-XANES is able to distinguish between $Cu^{2+}$ (CuO), $Cu^+$ ($Cu_2O$), $Cu_0$, and Cu alloy. The number of components per spectrum, and the amount of each component in it, was determined by principal component analysis and linear combination, respectively.

Before the start of the reaction (t = 0 min), Cu is present as an alloy. As soon as the catalyst comes into contact with the reaction mixture (100 ppm of aqueous nitrate solution), the alloy was converted into the segregated metallic copper phase. A small amount of $Cu^{2+}$ was also detected. The fraction of $Cu^{2+}$ present was governed by the flow of hydrogen, that is, at low hydrogen flow, there was more $Cu^{2+}$ and vice versa. Nitrate conversion also increased with the increase of hydrogen flow. When hydrogen was replaced by nitrogen, a steep increase of $Cu^{2+}$ fraction in detriment of alloy and copper metal phases was observed. Rapid formation of $Cu^{2+}$ fraction results in a drastic decrease in nitrate conversion. The activity could be partially recovered if the gas were switched back to hydrogen.

The experiments revealed that aqueous reduction of nitrate to nitrite over Pt-Cu catalyst involved the oxidation of copper sites to $Cu^{2+}$, which were consequently regenerated to the metallic or alloy via hydrogen spillover. Alloy and metallic copper phases were equally involved in the reduction process, whereas $Cu^{2+}$ phase was found unreactive. In the case of Pd-Cu catalyst, the activity was also dependent on hydrogen flow because it controls the availability of activated H groups in the surface; however, the deactivation process due to copper oxidation was slower due to the presence of Pt β-hydride phase. It should be mentioned that in the case of Pd-Cu, $Cu^+$ phase was also detected. The study validated retroactively the most consensual reaction mechanism.[68]

### 6.4.2.2  High-Pressure Hydrogenation Reactions

Nanosize gold has very peculiar catalytic properties. The motif of such unique properties has been a topic of intense debate in literature. One of the most discussed aspects is the oxidation state of gold during hydrogenation reactions, in particular in the hydrogenation of nitrobenzene. So far there is no consensus, with some authors arguing that cationic gold is essential for the process. Kartusch et al.[70] performed an operando study in which they monitored gold oxidation state and catalytic activity as a function of time on stream. They used $Au/CeO_2$, which was pretreated in hydrogen under different conditions so they could start the reaction with different concentrations of cationic gold in the catalyst. Figure 6.10 shows the outcome of the experiment. The experiments were performed in an autoclave reactor,[52] and Johann-type spectrometer.

As soon as the nitrobenzene was introduced, the concentration of cationic gold decreased, and eventually it disappeared per complete (after 10 min). With respect to activity, after 10 min the reaction rate decreased slightly, suggesting that cationic gold is in fact involved in the reaction, and somehow more active. However, similar behavior was observed with the catalyst that did not have any cationic gold from the start. The authors justify the decrease not on the basis of the oxidation state of gold but most likely due to surface poisoning by reaction intermediates. The experiments revealed not only that metallic gold is active in the reaction but also that cationic

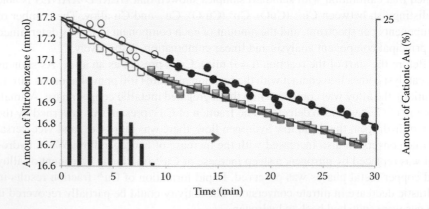

**FIGURE 6.10**  **(See Color Insert.)** Nitrobenzene hydrogenation over $Au/CeO_2$ catalysts with different concentrations of cationic gold ($Au^{3+}$). Bar plot represents $Au^{3+}$ fraction determined by Au $L_3$-edge HERFD-XANES, and points represent nitrobenzene conversion. (Red) $Au/CeO_2$ pretreated at 60°C, and (black) $Au/CeO_2$ pretreated at 100°C. Reaction carried out under 10 bar $H_2$, 100°C, and with a stirring rate of 1500 rpm.

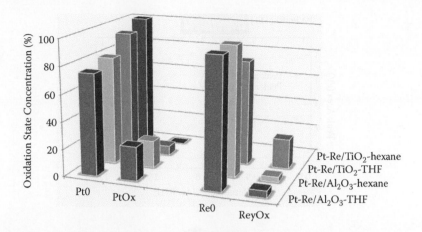

**FIGURE 6.11**  Average oxidation state of Pt and Re as a function of solvent and support. (Adapted from Sá, J. et al., 2011, *Chem. Commun.* 47, 6590–6592.[71] With permission.)

gold is not stable under 10 bar of hydrogen. Thus, cationic gold is not responsible for gold activity.

Sá et al.[71] used a similar experimental approach to determine the oxidation state of Pt and Re (4% Pt-4% Re) supported on $Al_2O_3$ or $TiO_2$ after liquid phase reduction in different solvents (THF or hexane). As it can be seen from Figure 6.11, upon reduction, the most predominant oxidation state of the metals was the metallic state but the amount of metallic phase in it depends on both support and solvent. This contrasts with gas phase reduction, where 100% of metals were reduced to metallic state. This reveals the importance of carrying out these experiments under working conditions.

### 6.4.3  ELECTROCHEMISTRY

#### 6.4.3.1  Electrochemical Oxygen Reduction

Proton exchange membrane fuel cells (PEMFC) are seen as a promising development for a renewable-energy infrastructure, which converts stored chemical energy into electricity.

An unaccomplished prerequisite is the development of cost-effective electrocatalysts, in particular for the oxygen reduction reaction (ORR). Electrocatalysts made of earth-abundant metals are desirable but they exhibit too high overpotentials.[72] Platinum-based catalysts show the best electrocatalytical abilities[73] and therefore significant effort has been made to understand the intricacies of the process over these systems.

*In situ* HERFD-XAS was used by Fribel et al.[74] to determine the effect of Pt geometric structure in the ORR reaction (Figure 6.12). When the experiments were carried out on a 1 monolayer of Pt on Rh(111) (2D Pt/Rh(111), they observed two distinct regions. At potentials below 1.0 eV the spectra are dominated by a Pt metal (peak 11566.5 eV) signal. At potentials higher than 1.0 eV, a new phase emerged, and its concentration increased with the potential, which was assigned to Pt(II) oxide (peak 11568.4 eV).[75]

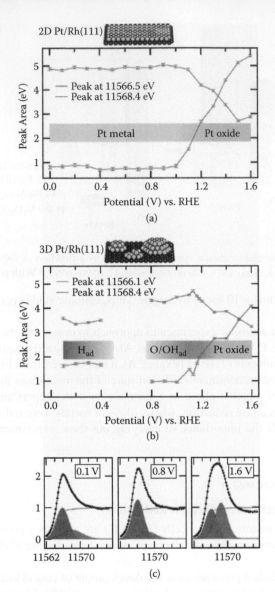

**FIGURE 6.12** **(See Color Insert.)** Deconvolution of *in situ* Pt L₃-edge HERFD-XAS for Pt/Rh(111) in 0.01 M HClO4. While only the Pt metal-to-oxide transition can be identified for the 2D Pt layer (a), the 3D deposit (b) shows additional spectral signatures due to H$_{ad}$ and O/OH$_{ad}$. Three representative fitting results from 3D Pt/Rh(111) are shown in (c) using the same colors as in (b) for the peak areas, and a gray line for the arctangent function. (Reprinted from Friebel et al., 2012, *J. Am. Chem. Soc.* 134, 9664–9671.[74] With permission. Copyright 2013, American Chemical Society.)

The same regions were observed when they used nanoparticles of Pt deposited on Rh(111) (3D Pt/Rh(111); however, in this case they were also able to resolve features associated to adsorbed molecules. At potentials close to hydrogen evolution (<0.4 eV) the whiteline intensity decreased and a shoulder at ~11570 eV appeared, assigned to chemisorbed hydrogen on platinum (Pt-H$_{ad}$). At potentials between 0.8 and 1.13 eV, the authors observed a small increase in the peak at 11566 eV without significant broadening ascribed to oxygen chemisorbed on platinum (Pt-O/OH$_{ad}$).

Similar results were observed when Pt was supported on activated carbon.[76] It should be mentioned that the detection of adsorbed species was only possible due to the use of high-energy resolution XAS. DFT calculations confirmed that the bond between atomic oxygen and 2D Pt/Rh(111) at 0.6 eV was weaker than on pure Pt surfaces, which is related to a shift of the $d$-band center to lower energy. 3D Pt/Rh(111) shows a high affinity to chemisorbed H at low potentials and O/OH at high potentials in comparison to 2D Pt/Rh(111). Hence, the $d$-band center was shifted back toward pure Pt, thus the increase of both Pt-O and Pt-H bond strength. This is a very important result because it opens the possibility to tune electrocatalyst's properties based on particle design instead of constituent elements.

Friebel et al.[77] performed *in situ* HERFD-XAS to elucidate the differences in Pt electronic structure when supported on different substrates, namely, Rh(111) and Au(111). In both cases, they observed a significant difference in the whiteline intensity, position, and broadness above 1.0 eV. These are signature changes related to the oxidation of platinum from metallic state to +2 (PtO). Spectral changes on Pt/Rh(111) ceased above 1.0 eV; however, in the case of Pt/Au(111) a further potential increase led to a strong increase in whiteline intensity and a blue edge shift of ca. 1 eV. The changes suggested the formation of PtO$_2$ (Pt(IV)), and its concentration increased when the potential was incremented. In analogy to a trend observed with $4d$ transition metals,[78] increased Pt cohesion in Pt/Au(111) is due to a shift in the $d$-band, which destabilizes subsurface oxygen (a precursor for oxide formation).

## 6.5  FINAL REMARKS

This chapter compiled a few studies in which the use of energy-resolved XAS and XES enabled access to heterogeneous catalysts, electronic states, which are responsible for the catalytic properties of any material. Most of the systems can now be studied under reaction conditions of temperature, pressure, and so on, which is often not the case with other spectroscopy techniques. An area that needs to be improved is time resolution. However, recent developments in dispersive spectrometers, pump-probe schemes, and high-flux synchrotrons and XFELs, suggest that in a short time one will be able to do experiments with femtosecond time resolution.

## REFERENCES

1. (a) Hammer, B. and J.K. Nørskov. Why gold is the noblest of all the metals. *Nature* 376 (1995) 238; (b) Blingaard, T., J.K. Nørskov, S. Dahl, J. Matthiesen, C.H. Christensen, and J. Shested. The Brønsted–Evans–Polanyi relation and the volcano curve in heterogeneous catalysis. *J. Catal.* 224 (2004) 206.

2. de Groot, F., G. Vankó, and P. Glatzel. The 1s X-ray absorption pre-edge structures in transition metal oxides. *J. Phys.: Condens. Matter* 21 (2009) 104207.

3. Kurian, R., M.M. van Schooneveld, N. Zoltán, G. Vankó, and F.M.F. de Groot. Temperature-dependent 1s2p resonant inelastic X-ray scattering of CoO. *J. Phys. Chem. C* 117 (2013) 2976.

4. de Groot, F. Multiple effects in X-ray spectroscopy. *Coord. Chem. Rev.* 249 (205) 31–56.

5. Szlachetko, J. and J. Sá. Rational design of oxynitride materials: From theory to experiment. *Cryst. Eng. Comm.* 15 (2013) 2583–2587.

6. (a) Hagen, J. *Industrial Catalysis: A Practical Approach*, 2nd ed., Wieheim: Wiley-VCH (2006); (b) Rase, H.F. *Handbook of Commercial Catalysts: Heterogeneous Catalysts*, Boca Raton, FL: CRC Press (2000).

7. (a) Russel, A.S. Passivity, catalytic action, and other phenomena. *Nature* 117 (1926) 47–48; (b) Roginsky, S. and E. Schulz. Katalytische Vorgänge in fester Phase. I. Die Zersetzung des Kaliumpermanganats. *Zeitschrift fue Physikalische Chemie-Stochiometrie und Verwandtschaftslehre* 138 (1928) 21–41.

8. Hammer, B. and J.K. Nørskov. Theoretical surface science and catalysis—Calculations and concepts. *Adv. Catal.* 45 (2000) 71–129.

9. Yu, W., M.D. Porosoff, and J.G. Chen. Review of Pt-based bimetallic catalysis: From model surfaces to supported catalysts. *Chem. Rev.* 112 (2012) 5780–5817.

10. (a) Lytle, F.W., P.S.P. Wei, R.B. Greegor, G.H. Via, and J.H. Sinfelt. Effect of chemical environment on magnitude of X-ray absorption resonance at $L_{III}$ edges. Studies on metallic elements, compounds, and catalysts. *J. Chem. Phys.* 70 (1979) 4849–4855; (b) L.F., Mattheiss, and R.E. Dietz. Relativistic tight-binding calculation of core-valence transitions in Pt and Au. *Phys. Rev. B: Condens. Matter* 22 (1980) 1663–1676.

11. Groot, F.M.F., M.H. Krisch, and J. Vogel. Spectral sharpening of the Pt L edges by high-resolution X-ray emission. *Phys. Rev. B: Condens. Matter* 66 (2002) 195112.

12. Anniyev, T., H. Ogasawara, M.P. Ljungberg, K.T. Wikfeldt, J.B. MacNaughton, L.-Å. Näslund, U. Bergmann, S. Koh, P. Strasser, L.G.M. Pettersson, and A. Nilson. Complementarity between high-energy photoelectron and L-edge spectroscopy for probing the electronic structure of 5d transition metalcatalysts. *Phys. Chem. Chem. Phys.* 12 (2010) 5694–5700.

13. (a) Haruta, M., T. Kobayashi, H. Sano, and N. Yamada. Novel gold catalysts for the oxidation of carbon monoxide at a temperature far below 0°C. *Chem. Lett.* 2 (1987) 405; (b) Valden, M., X. Lai, and D.W. Goodman. Onset of catalytic activity of gold clusters on titania with the appearance of nonmetallic properties. *Science* 281 (1998) 1647; (c) Mohr, C., H. Hofmeister, J. Radnik, and P. Claus. Identification of active sites in gold-catalyzed hydrogenation of acrolein. *J. Am. Chem. Soc.* 125 (2003) 1905; (d) Hughes, M.D., J.Y. Xu, P. Jenkins, P. McMorn, P. Landon, D.I. Enache, A.F. Carley, G.A. Attard, G.J. Hutchings, F. King, E.H. Stitt, P. Johnston, K. Griffin, and C.J. Kiely. Tunable gold catalysts for selective hydrocarbon oxidation under mild conditions. *Nature* 437 (2005) 1131.

14. (a) Mattheiss, L.F. and R.E. Dietz. Relativistic tight-binding calculation of core-valence transitions in Pt and Au. *Phys. Rev. B* 22 (1980) 1663; (b) Mason, M.G. Electronic strucuture of supported small metal clusters. *Phys. Rev. B* 27 (1983) 748; (c) Häkkinen, H., M. Moseler, and U. Landman. Bonding in Cu, Ag, and Au clusters: Relativistic

effects, trends, and surprises. *Phys. Rev. Lett.* 89 (2002) 033401; (d) Jain, P.K. A DFT-based study of the low-energy electronic structures and properties of small gold clusters. *Struct. Chem.* 16 (2005) 421.

15. Van Bokhoven, J.A. and J.T. Miller. van Bokhoven, J.A. and J.T. Miller. d-Electron density and energy as function of particle size for supported gold catalysts. *J. Phys. Chem. C* 111 (2007) 9245–9249.

16. (a) Farmer, J.A. and C.T. Campbell. Ceria maintains smaller metal catalyst particles by strong metal-support bonding. *Science* 329 (2010) 933–936; (b) Park, J.B., J. Graciani, J. Evans, D. Stacchiola, S.D. Senanayake, L. Barrio, P. Liu, J.F. Sanz, J. Hrbek, and J.A. Rodrigues. Gold, copper, and platinum nanoparticles dispersed on $CeO_x/TiO_2$ (110) Surfaces: High water-gas shift activity and the nature of the mixed-metal oxide at the nanometer level. *J. Am. Chem. Soc.* 132 (2009) 356–363.

17. (a) Steele, B.C.H. and A. Heinzel. Materials for fuel-cell technologies. *Nature* 414 (2001) 345–352; (b) Kosinski, M.R. and R.T. Baker. Preparation and property-performance relationships in samarium-doped ceria nanopowders for solid oxide fuel cell electrolytes. *J. Power Sources* 196 (2011) 2498–2512.

18. (a) Pierscionek, B.K., Y. Li, A. Yasseen, L.M. Colhoun, R.A. Schachar, and W. Chen. *Nanoceria* have no genotoxic effect on human lens epithelial cells. *Nanotechnology* 21 (2010) 035102; (b) Chen, J.P., S. Patil, S. Seal, and J.F. McGinnis. Rare earth nanoparticles prevent retinal degeneration induced by intracellular peroxides. *Nat. Nanotechnol.* 1 (2006) 142–150.

19. (a) Park, T., V.A. Sidorov, F. Ronning, J.X. Zhu, Y. Tokiwa, H. Lee, E.D. Bauer, R. Movshovich, J.L. Sarrao, and J.D. Thompson. Isotropic quantum scattering and unconventional superconductivity. *Nature* 456 (2008) 366–368; (b) Barla, A., J. Derr, J.P. Sanchez, B. Salce, G. Lapertot, B.P. Doyle, R. Ruffer, R. Lengsdorf, M.M. Abd-Elmeguid, and J. Flouquet. High-pressure ground state of $SmB_6$: Electronic conduction and long-range magnetic order. *Phys. Rev. Lett.* 94 (2005) 166401.

20. Babu, S., J.-H. Cho, J. Dowding, E. Heckert, C. Komanski, S. Das, J. Colon, C.H. Baker, M. Bass, W.T. Self, and S. Seal. Multicolored redox active upconverter cerium oxide nanoparticle for bio-imaging and therapeutics. *Chem. Commun.* 46 (2010) 6915–6917.

21. Oark, B., H. Li, and L.R. Corrales. Molecular dynamics simulation of $La_2O_3$-$Na_2O$-$SiO_2$ glasses. I. The structural role of $La^{3+}$ cations. *J. Non–Cryst. Solids* 297 (2002) 220–238.

22. (a) Motoyama, E.M., G. Yu, I.M. Vishik, O.P. Vajk, P.K. Mang, and M. Greven. Spin correlations in the electron-doped high-transition-temperature superconductor $Nd_{2-x}Ce_xCuO_{4\pm\delta}$. *Nature* 445 (2007) 186–189; (b) Nakatsuji, S., K. Kuga, Y. Machida, T. Tayama, T. Sakakibara, Y. Karaki, H. Ishimoto, S. Yonezawa, Y. Maeno, E. Pearson, G.G. Lonzarich, L. Balicas, H. Lee, and Z. Fisk. Superconductivity and quantum criticality in the heavy-fermion system β-$YbAlB_4$. *Nat. Phys.* 4 (2008) 603–607; (c) Dallera, C., M. Grioni, A. Shukla, G. Vanko, J.L. Sarrao, J.P. Rueff, and D.L. Cox. New spectroscopy solves an old puzzle: The Kondo Scale in heavy fermions. *Phys. Rev. Lett.* 88 (2002) 196403; (d) Tsujii, N., H. Kontani, and K. Yoshimura. Universality in heavy fermion systems with general degeneracy. *Phys. Rev. Lett.* 94 (2005) 057201.

23. (a) Fornasiero, P., G. Balducci, R. Di Monte, J. Kaspar, V. Sergo, G. Gubitosa, A. Ferrero, and M. Graziani. Modification of the redox behaviour of $CeO_2$ by structural doping with $ZrO_2$. *J. Catal.* 164 (1996) 173–183; (b) Shimizu, K., H. Kawachi, and A. Satsuma. Study of active sites and mechanism for soot oxidation by silver-loaded *Ceria* catalyst. *Appl. Catal.* B 96 (2010) 169–175.

24. Shao, Z., S.M. Haile, J. Ahn, P.D. Ronney, Z. Zhan, and S.A. Barnett. A thermally self-sustained micro solid-oxide fuel-cell stack with high power density. *Nature* 435 (2005) 795–798.

25. Fu, Q., H. Saltsburg, and M. Flytzani-Stephanopoulos. Active nonmetallic Au and Pt species on *Ceria*-based water-gas shift catalysts. *Science* 301 (2003) 935–938.

26. Trovarelli, A. *Catalysis by Ceria and Related Materials*, G.J. Hutchings, ed., Catalysis Science Series, London: Imperial College Press (2002).

27. Paun, C., O.V. Safonova, J. Szlachetko, P.M. Abdala, M. Nachtegaal, J. Sá, E. Kleymenov, A. Cervelino, F. Krumeich, and J.A. van Bokhoven. Polyhedral $CeO_2$ nanoparticles: Size-dependent geometrical and electronic structure. *J. Phys. Chem. C* 116 (2012) 7312–7317.

28. (a) Hämäläinen, K., D.P. Siddons, J.B. Hastings, and L.E. Berman. Elimination of inner-shell lifetime broadening in X-ray-absorption spectroscopy. *Phys. Rev. Lett.* 7 (1991) 2850–2853; (b) Kvashnina, K.O., S.M. Butorin, and P. Glatzel. Direct study of the f-electron configuration in lanthanide systems. *J. Anal. Atmos. Spectrom.* 26 (2011) 1265–1272; (c) Kotani, A., K.O. Kvashnina, S.M. Butorin, and P. Glatzel. A new method of directly determining the core-hole effect in the Ce $L_3$ XAS of mixed valence Ce compounds—An application of resonant X-ray emission spectroscopy. *J. Electron Spectrosc. Relat. Phenom.* 184 (2011) 210–215; (d) Soldatov, A.V., T.S. Ivanchenko, S. Dellalonga, A. Kotani, Y. Iwamoto, and A. Bianconi. Crystal-strucuture effects in the Ce $L_3$-edge X-ray-absorption spectrum of $CeO_2$: Multiple-scattering resonances and many-body final states. *Phys. Rev. B* 50 (1994) 5074–5080.

29. Somorjai, G.A. and C. Aliaga. Molecular studies of model surfaces of metals from single crystals to nanoparticles under catalytic reaction conditions. Evolution from prenatal and postmortem studies of catalysts. *Langmuir* 26 (2010) 16190–16203.

30. Hammer, B. and J.K. Nørskov. Electronic factors determining the reactivity of metal surfaces. *Surf. Sci.* 343 (1995) 211.

31. (a) (IPA) *International Platinum Group Metals Association*, http://www.ipa-news.com/en/89-0-Catalytic+Converters.html; (b) Farrauto, R.J. and R.M. Heck. Catalytic converters: State of the art and perspectives. *Catal. Today* 51 (1999) 351.

32. (a) Trimm, D.L. and Z.I. Önsan. Onboard fuel conversion for hydrogen-fuel-cell-driven vehicles. *Catal. Rev.* 43 (2001) 31–84; (b) Jacobson, M.Z., W.G. Colella, and D.M. Golden. Cleaning the air and improving health with hydrogen fuel-cell vehicles. *Science* 308 (2005) 1901–1905.

33. (a) Ertl, G., M. Neumann, and K.M. Streit. Chemisorption of CO on the Pt (111) surface. *Surf. Scie.* 64 (1977) 393–410; (b) Gritsch, T., D. Coulman, R.J. Behm, and G. Ertl, Mechanism of the CO-induced *1x2–1x1* structural transformation of Pt(110). *Phys. Rev. Lett.* 63 (1989) 1086–1089; (c) Krischer, K., M. Eiswirth, and G. Ertl. Oscillatory CO oxidation on Pt(110): Modeling of temporal self-organization. *J. Chem. Phys.* 96 (1992) 9161.

34. Safonova, O.V., M. Tromp, J.A. van Bokhoven, F.M.F. de Groot, J. Evans, and P. Glatzel. Identification of CO adsorption sites in supported Pt catalysts using high-energy-resolution fluorescence detection X-ray spectroscopy. *J. Phys. Chem. B* 110 (2006) 16162–16164.

35. (a) Ankudinov, A.L., B. Ravel, J.J. Rehr, and S.D. Conradson. Real-space multiple-scattering calculation and interpretation of X-ray-absorption near-edge structure. *Phys. Rev B* 58 (1998) 7565; (b) Rehr, J.J. and R.C. Albers, Theoretical approaches to X-ray absorption fine structure. *Rev. Mod. Phys.* 72 (2000) 621; (c) Ankudinov, A.L., J.J. Rehr, J. Low, and S.R. Bare. Effect of hydrogen adsorption on the X-ray absorption spectra of small Pt clusters. *Phys. Rev. Lett.* 86 (2001) 1642.

36. (a) Glatzel, P., J. Singh, K.O. Kvashnina, and J.A. van Bokhoven. *In situ* characterization of 5d density of states of Pt nanoparticles upon adsorption of CO. *J. Am. Chem. Soc.* 132 (2010) 2555–2557; (b) Singh, J., R.C. Nelson, B.C. Vicente, S.L. Scott, and J.A. van Bokhoven. Electronic structure of alumina-supported monometallic Pt and bimetallic PtSn catalysts under hydrogen and carbon monoxide environment. *Phys. Chem. Chem. Phys.* 12 (2010) 5668–5677.

37. (a) Blyholder, G. Molecular orbital view of chemisorbed carbon monoxide. *J. Phys. Chem.* 68 (1964) 2772; (b) Chen, L., B. Chen, C. Zhou, J. Wu, R.C. Forrey, and H. Cheng. Influence of CO poisoning on hydrogen chemisorption onto a $Pt_6$ cluster. *J. Phys. Chem. C* 112 (2008) 13937; (c) Valero, M.C., P. Raybaud, and P. Sautet. Interplay between molecular adsorption and metal–support interaction for small supported metal clusters: CO and $C_2H_4$ adsorption on $Pd_4/\gamma$ -$Al_2O_3$. *J. Catal.* 247 (2007) 339.

38. Ertl, G., H. Knözinger, F. Schüth, and J. Weitkamp, eds. *Handbook of Heterogeneous Catalysis*, vol. 8, 2nd ed., Weinheim, Germany: Wiley-VCH (2008).

39. Köningsberger, D.C. and J.W. Cook Jr. *EXAFS and Near Edge Structure*, A. Bianconi, L. Incoccia, and S. Stipcich, eds., Berlin, Germany: Springer-Verlag (1983) p. 412.

40. Sá, J., J. Szlachetko, M. Sikora, M. Kavčič, O.V. Safonova, and M. Nachtegaal. Magnetic manipulation of molecules on a non-magnetic catalytic surface. *Nanoscale* 5 (2013) 8462.

41. (a) Liu, Z.-P. and P. Hu, CO oxidation and NO reduction on metal surfaces: Density functional theory investigations. *Top. Catal.* 28 (2004) 71–78; (b) Zhang, C.J. and P. Hu. CO oxidation on Pd(100) and Pd(111): A comparative study of reaction pathways and reactivity at low and medium coverages. *J. Am. Chem. Soc.* 123 (2001) 1166–1172.

42. Orita, H., N. Itoh, and Y. Inada. All electron scalar relativistic calculations on adsorption of CO on Pt(111) with full-geometry optimization: A correct estimation for CO site-preference. *Chem. Phys. Lett.* 384 (2004) 271–276.

43. (a) Clausen, B.S., L. Gråbæk, G. Steffensen, P.L. Hansen, and H. Topsøe. A combined QEXAFS/XRD method for on-line, *in situ* studies of catalysts: Examples of dynamic measurements of Cu-based methanol catalysts. *Catal. Lett.* 20 (1993) 23; (b) Couves, J.W., J.M. Thomas, D. Waller, R.H. Jones, A.J. Dent, G.E. Derbyshire, and G.N. Greaves. Tracing the conversion of aurichalcite to a copper catalyst by combined X-ray absorption and diffraction. *Nature* 354 (1991) 465; (c) Tsakoumis, N.E., A. Voronov, M. Rønning, W. van Beek, Ø. Borg, E. Rytter, and A. Holmen. Fischer–Tropsch synthesis: An XAS/XRPD combined *in situ* study from catalyst activation to deactivation. *J. Catal.* 291 (2012) 138–148.

44. Vicente, B.C., R.C. Nelson, J. Singh, S.L. Scott, and J.A. van Bokhoven. Electronic structures of supported Pt and PtSn nanoparticles in the presence of adsorbates and during CO oxidation. *Catal. Today* 160 (2011) 137–143.

45. (a) Panov, G.I., G.A. Sheveleva, A.S. Kharitonov, V.N. Romannikov, and L.A. Vostrikova. Oxidation of benzene to phenol by nitrous oxide over Fe-ZSM-5 zeolites. *Appl. Catal. A* 82 (1992) 31; (b) Panov, G.I., A.S. Kharitonov, and V.I. Sobolev. Oxidative hydroxylation using dinitrogen monoxide: A possible route for organic synthesis over zeolytes. *Appl. Catal. A* 98 (1993) 1.

46. Dubkov, K.A., N.S. Ovanesyan, A.A. Shteinman, E.V. Starokon, and G.I. Panov. Evolution of iron states and formation of α-sites upon activation of FeZSM-5 zeolites. *J. Catal.* 207 (2002) 341.

47. Kiwi-Minsker, L., D.A. Bulushev, and A. Renken. Active sites in HZSM-5 with low Fe content for the formation of surface oxygen by decomposing $N_2O$: Is every deposited oxygen active? *J. Catal.* 219 (2003) 273.

48. Pirnbruber, G.D., J.-D. Grunwaldt, P.K. Roy, J.A. van Bokhoven, O. Safonova, and P. Glatzel. The nature of the active site in the Fe-ZSM-5/$N_2O$ system studied by (resonant) inelastic X-ray scattering. *Catal. Today* 126 (2007) 127–134.

49. Overweg, A.R., M.W.J. Craje, A.M. van der Kraan, I. Arends, A. Ribera, and R.A. Sheldon. Remarkable $N_2$ affinity of a steam-activated FeZSM-5 catalyst: A $^{57}Fe$ Mössbauer study. *J. Catal.* 223 (2004) 262.

50. Manyar, H.G., R. Morgan, K. Morgan, B. Yang, P. Hu, J. Szlachetko, J. Sá, and C. Hardacre. High energy resolution fluorescence detection XANES—An *in situ* method to study the interaction of adsorbed molecules with metal catalysts in the liquid phase. *Catal. Sci. Technol.* 3 (2013) 1497–1500.

51. Manyar, H.G., B. Yang, H. Daly, H. Moor, S. McMonagle, Y. Tao, G.D. Yadav, A. Goguet, P. Hu, and C. Hardacre. Selective hydrogenation of alpha-beta-unsaturated aldehydes and ketones using novel manganese oxide and platinum supported on manganese oxide octahedral molecular sieves as catalysts. *Chem. Cat. Chem.* 5 (2013) 506–512.

52. Makosch, M., C. Kartusch, J. Sá, R.B. Duarte, J.A. van Bokhoven, K. Kvashnina, P. Glatzel, J. Szlachetko, D.L.A. Fernandes, E. Kleymenov, M. Nachtegaal, and K. Hungerbühler. HERFD XAS/ATR-FTIR batch reactor cell. *Phys. Chem. Chem. Phys.* 14 (2012) 2164–2170.

53. (a) Sanfilippo, D. Dehydrogenation of paraffins: Key technology for petrochemicals and fuels. *Cattech* 4 (2000) 56–73; (b) Akporiaye, D., S.F. Jensen, U. Olsbye, F. Rohr, E. Rytter, M. Ronnekleiv, A.I. Spjelkavik, and A. Novel. A novel, highly efficient catalyst for propane dehydrogenation. *Ind. Eng. Chem. Res.* 40 (2001) 4741–4748.

54. Bruch, R. and A.J. Mitchell. The role of tin and rhenium in bimetallic reforming catalysts. *Appl. Catal.* 6 (1083) 121–128.

55. (a) Bariås, O.A., A. Holmen, E.A. Blekkan. Propane dehydrogenation over supported platinum catalysts: Effect of tin as a promoter. *Catal. Today* 24 (1995) 361–364; (b) Llorca, J., N. Homs, J. León, J. Sales, J.L.G. Fierro, and P. Ramirez de la Piscina. Supported Pt–Sn catalysts highly selective for isobutane dehydrogenation: Preparation, characterization and catalytic behavior. *Appl. Catal.* A 189 (1999) 77–86.

56. Iglesias-Juez, A., A.M. Beale, K. Maaijen, T.C. Weng, P. Glatzel, and B.M. Weckhuysen, A combined in situ time-resolved UV–Vis, Raman and high-energy resolution X-ray absorption spectroscopy study on the deactivation behavior of Pt and Pt-Sn propane dehydrogenation catalysts under industrial reaction conditions. *J. Catal.* 276 (2010) 268–279.

57. Chinchen, G., P. Denny, J. Jennings, M. Spencer, and K. Waugh. Synthesis of methanol: Part 1. Catalysts and kinetics. *Appl. Catal.* 36 (1998) 1–65.

58. Kleymenov, E., J. Sá, J. Abu-Dahrieh, D. Rooney, J.A. van Bokhoven, E. Troussard, J. Szlachetko, O.V. Safonova, and M. Nachtegaal. Structure of the methanol synthesis catalyst determined by *in situ* HERFD XAS and EXAFS. *Catal. Sci. Technol.* 2 (2012) 373–378.

59. (a) Topsoe, N.-Y. and H. Topsoe. On the nature of surface structural changes in Cu/ZnO methanol synthesis catalysts. *Top. Catal.* 8 (1999) 267–270; (b) Fujitani, T. and J. Nakamura. The effect of ZnO in methanol synthesis catalysts on Cu dispersion and the specific activity. *Catal. Lett.* 56 (1998) 119–124.

60. Groothaert, M.H., P.J. Smeets, B.F. Sels, O.A. Jacobs, and R.A. Schoonheydt. Selective oxidation of methane by the Bis($\mu$-oxo)dicopper core stabilized on ZSM-5 and mordenite zeolytes. *J. Am. Chem. Soc.* 127 (2005) 1394–1395.

61. (a) Beznis, N.V., B.M. Weckhuysen, and J.H. Bitter. Cu-ZSM-5 zeolites for the formation of methanol from methane and oxygen: Probing the active sites and spectator species. *Catal. Lett.* 138 (2010) 14–22; (b) Dubkov, K.A., V.I. Sobolev, and G.I. Panov. Low-temperature oxidation of methane to nethanol on FeZSM-5 zeolite. *Kinet. Catal.* 39 (1998) 72–79.

62. Alayon, E.M., C.M. Nactegaal, E. Kleymenov, and J.A. van Bokhoven. Determination of the electronic and geometric structure of Cu sites during methane conversion over Cu-MOR with X-ray absorption spectroscopy. *Micro. Meso. Mater.* 166 (2013) 131–136.

63. Lytle, F.W. and R.R. Greegor. Discussion of X-ray-absorption near-edge structure: Application to Cu in the high-$Tc$ superconductors $La_{1.8}Sr_{0.2}CuO_4$ and $YBa_2Cu_3O_7$. *Phys. Rev. B* 37 (1988) 1550–1562.

64. (a) Haruta, M., N. Yamada, T. Kobayashi, and S. Ijima. Gold catalysts prepared by coprecipitation for low-temperature oxidation of hydrogen and of carbon monoxide. *J. Catal.* 115 (1989) 301; (b) Valden, M., X. Lai, and D.W. Goodman. Onset of catalytic activity of gold clusters on *Titania* with the appearance of nonmetallic properties. *Science* 281 (1998) 1647.

65. (a) Chen, M.S. and D.W. Goodman. The structure of catalytically active gold on *Titania*. *Science* 306 (2004) 252–255; (b) Campbell, C.T. The active site in nanoparticle gold catalysis. *Science* 306 (2004) 234–235.

66. Van Bokhoven, J.A., C. Louis, J.T. Miller, M. Tromp, O.V. Safonova, and P. Glatzel. Activation of oxygen on gold/alumina catalysts: *In situ* high-energy-resolution fluorescence and time-resolved X-ray spectroscopy. *Angew. Chem.* 118 (2006) 4767–4770.

67. Wikler, C., A.J. Carew, S. Haq, and R. Raval. Carbon monoxide on γ-alumina single crystal surfaces with gold nanoparticles. *Langmuir* 19 (2003) 717.

68. Barrabés, N. and J. Sá. Catalytic nitrate removal from water, past, present and future perspectives. *Appl. Catal.* B 104 (2011) 1–5.

69. Sá, J., N. Barrabés, E. Kleymenov, C. Lin, K. Föttinger, O.V. Safonova, J. Szlachetko, J.A. van Bokhoven, M. Nachtegaal, A. Urakawa, G.A. Crespo, and G. Rupprechter. The oxidation state of copper in bimetallic (Pt-Cu, Pd-Cu) catalysts during water denitration. *Catal. Sci. Technol.* 2 (2012) 794–799.

70. Kartusch, C., M. Makosch, J. Sá, K. Hungerbuehler, and J.A. van Bokhoven. The dynamic structure of gold supported on *Ceria* in the liquid phase hydrogenation of nitrobenzene. *Chem Cat Chem* 4 (2012) 236.

71. Sá, J., C. Kartusch, M. Makosch, C. Paun, J.A. van Bokhoven, E. Kleymenov, J. Szlachetko, M. Nachtegaal, H.G. Manyar, and C. Hardacre. Evaluation of Pt and Re oxidation state in a pressurized reactor: difference in reduction between gas and liquid phase. *Chem. Commun.* 47 (2011) 6590–6592.

72. (a) Bashyam, R. and P. Zelenay. A class of non-precious metal composite catalysts for fuel cells. *Nature* 443 (2006) 63–66; (b) Blizanac, B.B., P.N. Ross, N.M. Markovic. Oxygen electroreduction on Ag(1 1 1): The pH effect. *Electrochim. Acta* 52 (2007) 2264–2271.

73. (a) Strasser, P., S. Koh, T. Anniyev, J. Greeley, K. More, C. Yu, Z. Liu, S. Kaya, D. Nordlund, H. Ogasawara, M.F. Toney, and A. Nilson. Lattice-strain control of the activity in dealloyed core-shell fuel cell catalysts. *Nat. Chem.* 2 (2010) 454–460; (b) Zhang, J.L., M.B. Vukmirovic, M. Mavrikakis, and R.R. Adzic. Controlling the catalytic activity of platinum-monolayer electrocatalysts for oxygen reduction with different substrates. *Angew. Chem. Int. Ed.* 44 (2005) 2132–2135; (c) Greeley, J., I.E.L. Stephens, A.S. Bondarenko, T.P. Johansson, H.A. Hansen, T.F. Jaramillo, J. Rossmeisl, I. Chrokendorff, and J.K. Nørskov. Alloys of platinum and early transition metals as oxygen reduction electrocatalysts. *Nat. Chem.* 1 (2009) 552–556; (d) Zhang, J., K. Sasaki, E. Sutter, and R.R. Adzic. Stabilization of platinum oxygen-reduction electrocatalysts using gold clusters. *Science* 315 (2007) 220–222; (e) Stamenkovic, V.R., B. Fowler, B.S. Mun, G.F. Wang, P.N. Ross, C.A. Lucas, and N.M. Markovic. Improved oxygen reduction activity on Pt$_3$Ni(111) via increased surface site availability. *Science* 315 (2007) 493–497; (f) Stamenkovic, V.R., B.S. Mun, M. Arenz, K.J.J. Mayrhofer, C.A. Lucas, G.F. Wang, P.N. Ross, and N.M. Markovic. Trends in electrocatalysis on extended and nanoscale Pt-bimetallic alloy surface. *Nat. Mater.* 6 (2007) 241–247.

74. Friebel, D., V. Viswanathan, D.J. Miller, T. Anniyev, H. Ogasawara, A.H. Larsen, C.P. O'Grady, J.K. Nørskov, and A. Nilsson. Balance of nanostructure and bimetallic interactions in Pt model fuel cell catalysts: *In situ* XAS and DFT study. *J. Am. Chem. Soc.* 134 (2012) 9664–9671.

75. Friebel, D., D.J. Miller, C.P. O'Grady, T. Anniyev, J. Bargar, U. Nergmann, H. Ogasawara, K.T. Wikfeldt, L.G.M. Pettersson, and A. Nilsson. *In situ* X-ray probing reveals fingerprints of surface platinum oxide. *Phys. Chem. Chem. Phys.* 13 (2011) 262.

76. Merte, L.R., F. Bahafarid, D.J. Miller, D. Friebel, S. Cho, F. Mbuga, D. Sokaras, R. Aloso-Mori, T.-C. Weng, D. Nordlund, A. Nilsson, and B.R. Cuenya. Electrochemical oxidation of size-selected Pt nanoparticles studied using *in situ* high-energy-resolution X-ray absorption apectroscopy. *ACS Catal.* 2 (2012) 2371–2376.

77. Friebel, D., D.J. Miller, D. Nordlund, H. Ogasawara, and A. Nilsson. Degradation of bimetallic model electrocatalysts: An *in situ* X-ray absorption spectroscopy study. *Angew. Chem. Int.* Ed. 50 (2011) 10190–10192.

78. Todorova, M., W.X. Li, M.V. Ganduglia-Pirovano, C. Stampfl, K. Reuter, and M. Scheffler. Role of subsurface oxygen in oxide formation at transition metal surfaces. *Phys. Rev. Lett.* 89 (2002) 096103.

# Index

Printed and bound by CPI Group (UK) Ltd, Croydon, CR0 4YY

18/10/2024

01776208-0006